Synergy of Bio-Chemical Processes for Photocatalytic and Photoelectrochemical Wastewater Treatment

Scrivener Publishing
100 Cummings Center, Suite 541J
Beverly, MA 01915-6106

Publishers at Scrivener
Martin Scrivener (martin@scrivenerpublishing.com)
Phillip Carmical (pcarmical@scrivenerpublishing.com)

Synergy of Bio-Chemical Processes for Photocatalytic and Photoelectrochemical Wastewater Treatment

Edited by

Sadanand Pandey
Faculty of Applied Sciences and Biotechnology, Shoolini University, India

Elvis Fosso-Kankeu
Department of Metallurgy at the University of Johannesburg, South Africa

and

Soumya Pandit
Dept. of Biotechnology, Sharda University, Greater Noida, Delhi, India

Scrivener Publishing

WILEY

This edition first published 2024 by John Wiley & Sons, Inc., 111 River Street, Hoboken, NJ 07030, USA
and Scrivener Publishing LLC, 100 Cummings Center, Suite 541J, Beverly, MA 01915, USA

For more information about Scrivener publications please visit www.scrivenerpublishing.com.

Wiley Global Headquarters
111 River Street, Hoboken, NJ 07030, USA

For details of our global editorial offices, customer services, and more information about Wiley products visit us at www.wiley.com.

Library of Congress Cataloging-in-Publication Data

ISBN 978-1-394-19787-3

Cover image: Pixabay.Com
Cover design by Russell Richardson

Set in size of 11pt and Minion Pro by Manila Typesetting Company, Makati, Philippines

Printed in the USA

10 9 8 7 6 5 4 3 2 1

Contents

Preface

With the increase of recalcitrant organic pollutants in industrial wastewater, the need for a sustainable bio-electrochemical process has become pressing in order to ensure that treatment processes are coupled with some beneficiation advantages. Microbial fuel cells combine wastewater treatment and biological power generation. However, the resistance of these organic pollutants to biological degradation requires further adjustment of the system to improve sustainability through maximization of energy production.

Solar energy conversion using photocatalysis has drawn huge attention for its potential to provide renewable and sustainable energy. Furthermore, it might be the solution to serious environmental and energy-related problems. It has been widely understood for several years that the top global issues today are concerned with securing a clean supply of water and ensuring a reasonable price for clean energy. Researchers are studying advanced materials and processes to produce clean, renewable hydrogen fuel through photocatalytic and photoelectrocatalytic water splitting, as well as to reduce carbon dioxide from the air into fuels through photocatalysis. Limited progress is occurring in these areas.

Coupling photocatalysis function with fuel cell operation has been proposed by several authors as an alternative technology—namely photocatalysis fuel cells (PFCs). This has resulted in an increased rate of the reactions, as well as enhanced cathode electron transfer.

The purpose of this book is to comprehensively cover the evolvement in the conceptualization and application of photocatalytic fuel cells, as well as make a critical assessment of the contribution in the field of sustainable wastewater treatment and renewable energy production.

This book contains nine specialized chapters that provide comprehensive coverage of the design of photocatalytic fuel cells and their applications, including environmental remediation, chemical synthesis, green energy generation, model simulation for scaling up processes and implementation, and most importantly maximization of hydrogen evolution, recovery and applications.

A profound and immediate interest lies in the coupling of photocatalysis function and fuel cell operation with a set of unique and exclusive characteristics, including high catalytic activity, impressive selectivity, long-term durability, and environmental sustainability. The particularity of this book is that it synchronizes the potential of photocatalysis and microbial fuel cells for hydrogen recovery as a valuable energy source.

This book will be of interest to academic researchers and graduate students from the fields of environment, chemistry, engineering, and professionals that include environmental managers in the industries, water treatment plants managers and operators, water authorities, government regulatory bodies officers, and environmentalists.

The editors and the publisher are grateful to the reviewers who have contributed to improving the quality of the book through their constructive comments. The editors also thank Scrivener Publishing for including this book in their series.

Elvis Fosso-Kankeu
Sadanand Pandey
Soumya Pandit
February 2024

1

A Comprehensive Review of Graphitic Carbon Nitride (GCN/GC_3N_4) as a Promising Photocatalyst

Pragna M. Shivannavar[1], Chiranth Srirangapatna Puttasrinivasa[1], Sharmila Suresh[1], Charan Kumar Kachintaya[2], Lingaraju Honnur Gurusiddappa[3], Supreeth Mohan Kumar[4] and Shankramma Kalikeri[1]*

[1]Division of Nanoscience and Technology, School of Lifesciences, JSS Academy of Higher Education and Research, Sri Shivarathreeshwara Nagara, Mysuru, Karnataka, India

[2]#9A, Shankar Colony I Cross, SN Pet, Ballari, India

[3]Department of Environmental Science, School of Lifesciences, JSS Academy of Higher Education and Research, Sri Shivarathreeshwara Nagara, Mysuru, Karnataka, India

[4]Department of Microbiology, School of Life Sciences, JSS Academy of Higher Education and Research, Sri Shivarathreeshwara Nagara, Mysuru, Karnataka, India

Abstract

Over the last few decades, there has been a surge in interest in solar energy utilization due to the growing awareness of environmental protection and energy conservation. Photocatalysis has developed as one of the most crating tools for removal of pollutants in water and energy production. Due to the large bandgap and charge recombination, researchers are still struggling to meet the requirements of energy conversion efficacy. As a result, in the search for an active photocatalyst, graphitic carbon nitride (gC_3N_4/GCN), an active polymeric semiconductor, has sparked a vanguard of interest in the next generation of researchers. We highlighted crucial challenges associated with GCN technologies and severe opportunities in the aspect of developing upcoming groups of GCN-based materials in environmental fields. We also pointed the future directions and focused on GCN and GCN-based

**Corresponding author*: shankrutk@jssuni.edu.in; shankrutk@gmail.com

Sadanand Pandey, Elvis Fosso-Kankeu and Soumya Pandit (eds.) Synergy of Bio-Chemical Processes for Photocatalytic and Photoelectrochemical Wastewater Treatment, (1–26) © 2024 Scrivener Publishing LLC

advanced functional materials. Hopefully, this review can establish a link between the newly developed GCN materials and actual requirements for future commercial applications.

Keywords: Graphitic carbon nitride (gC_3N_4), visible light, wastewater treatment, sustainability

1.1 Introduction

Water is an important resource for humans as it is used for cleaning, drinking, cooking, and many more daily needs. The growth in population and industrialization has become the reason for the scarcity of clean water. Wastewater treatment has been gaining great interest to obtain clean water, and many ways have been used for the water treatment. The industries that manufacture paper, textiles, tanneries, photography, and food use azo dyes the most [1–5]. They are thought to be non-biodegradable in aerobic circumstances and are hazardous to the environment. Because the azo dye precursors are carcinogenic and immediately impair human health when consumed in water without any prior treatment, they have a significant negative environmental impact [1–3]. Conventional methods of wastewater treatment include membrane filtration [6, 7], "activated sludge process" [8, 9], coagulation, sedimentation, reverse osmosis. Membrane filtration is the frequently used conventional method because of its easier operation and effortless synthesis, but they also have few drawbacks such as membrane material instability durability, fouling resistance, and lesser water flux [10–12]. Also, the other conventional methods that are used have few drawbacks such as they do not remove the hazardous wastes completely and are time-consuming [13]. Thus, other advanced methods are needed to be developed for efficient treatment of the wastewater.

Advanced oxidation process (AOP) is a type of advanced method where chemical treatment is done to remove organic pollutants through oxidation reactions with hydroxyl radicals [5]. AOP is an efficient and developing method with the application of new apparatus, and also they do not have any secondary products at the end of the process [14, 15]. Photocatalysis is one such AOP that is gaining more interest as a method for wastewater treatment. This process results in the formation of benign products distinctive from the conventional methods where the pollutants are passed from one medium to the other [5]. The advantages of photocatalysis method includes mild operating conditions, cost effective, no secondary pollutant generation, less time, lesser chemical input, and eco-friendly [16, 17].

Photocatalyst is a key character for the efficiency of the photocatalytic process. An ideal photocatalyst used for photocatalysis must be sustainable, visible-light–driven [5, 18], environmentally friendly, economically feasible, and highly stable [19]. Semiconductors are generally used as photocatalyst. They may be metal-based inorganic photocatalysts such as TaON [20], $BiVO_4$ [21], CdS [22, 23], Ta_3N_5 [24, 25], $ZnlnS_4$, and $CulnS_2$ [26, 27] as well as metal oxide photocatalysts such as BiOCl [28], Ag_3PO_4 [29], TiO_2 [30, 31], WO_3 [32], and BiOBr [33]. These metal nanoparticles (NPs) also have limitations, such as lower efficiency, instability, higher costs, environmental toxicity, and its photo-corrosion, which are the causes for its unsuitability as photocatalysts [34, 35].

To overcome these issues, non-metal photocatalysts are developed, which consists of earth abundant elements such as carbon, sulfur, nitrogen, and phosphorous [36, 37]. Organic polymeric photocatalysts are among the non-metallic photocatalysts that is gaining importance these days for their wide applications. GCN is one such graphene based photocatalyst, with enhanced surface area and greater light harvesting efficiency, which increases its photocatalytic activity, and GCN-based photocatalyst is discussed in this chapter. Also, this chapter gives an insight of the modification of GCN to improve their photocatalytic activity.

1.2 GCN as a Photocatalyst

GCN is a non-metal polymeric photocatalyst that is gaining great interest from research perspective in the field of semiconductor photocatalysts. It is an n-type semiconductor. Allotropes of carbon nitride (C_3N_4) are α-C_3N_4, β-C_3N_4, and g-C_3N_4, pseudo-cubic and cubic-C_3N_4, of which, under ambient condition, g-C_3N_4 is a stable allotropic form [38, 39]. The building blocks of GCN are tri-s-triazine (C_6N_7) and triazine (C_3N_3) units that are depicted in Figure 1.1 [40, 41]. Observations have been made that "C_6N_7 units are more stable than the C_3N_3 units" [42]. It is composed of nitrogen (N) and carbon (C) that are piled up in layers of tri-s-triazine linked through amino groups [43]. Figure 1.2 shows the polymerized GCN [44]; GCN is also called as "melon" and has gained importance as photocatalyst because of its easy procedure of synthesis, better thermal and chemical stability, low cost, and enhanced visible-light absorption [45–47]. The valence band and conduction band of GCN are positioned at −1.1 eV and +1.6 eV, respectively, which makes the bandgap of GCN to be 2.7eV [48]. The bonds present between the layers, i.e., the van der Waals force of attraction gives GCN its chemical stability [48]. The color of GCN ranges from light

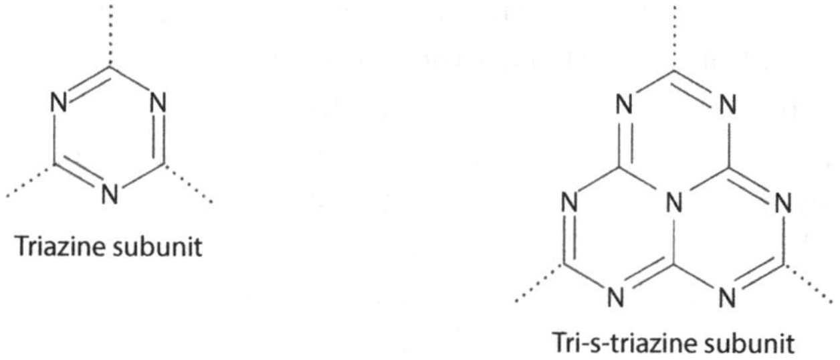

Figure 1.1 Triazine and tri-s-triazine units.

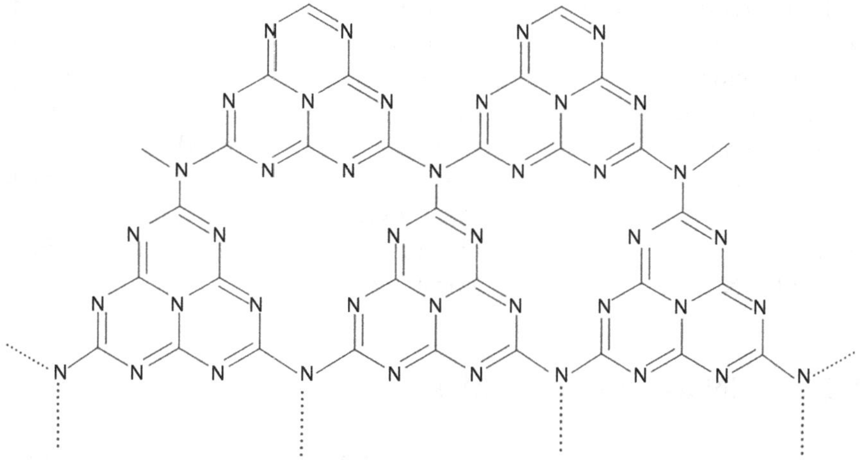

Figure 1.2 Polymerized structure of GCN.

yellow to brownish [49, 50]. Figure 1.3 shows "the photocatalytic mechanism for mineralization of the dyes"; as shown in the figure, the production of signals (charge carrier) is the indication for the start of the photocatalytic reaction; when GCN is illuminated with enough energy, "the photon's energy is transferred to the electrons in the valence band, which are then excited to the conduction band and produce holes in the valence band." The electron–hole pair then moves to the surface of GCN, where the holes in the valence band are consumed by the oxidizing electron donors, and the "protons are reduced to hydrogen by the electrons in the conduction band." Table 1.1 shows the GCN nanoparticles for dye degradation and other applications.

According to J. Xu *et al.* (2013), single atomic layer GCN nanosheets are fabricated using the chemical exfoliation process with dicyanamide

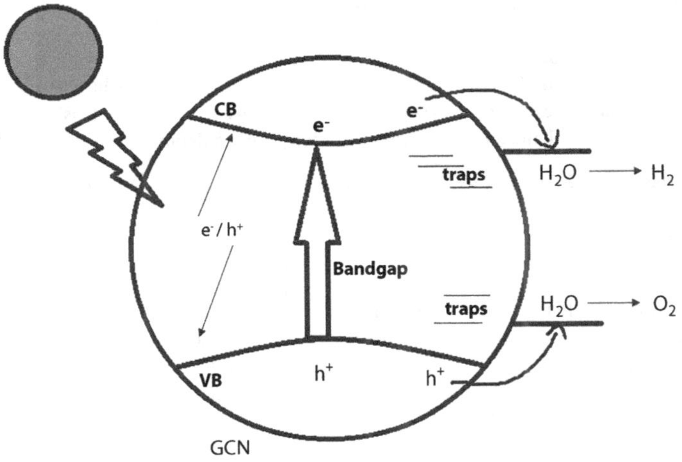

Figure 1.3 Photocatalysis mechanism of dye.

as a precursor, which results in single atomic layer GCN nanosheets with greater intrinsic properties. They utilized 25 mg of the GCN to accelerate under artificial visible light (500-W Xenon lamp) for degradation of MB, and it was found that 94% of the dye was degraded within 4 h. "It is generally recognized that high surface area, light absorption, and charge transport have a greater influence on photocatalytic activity."

Li *et al.* (2015) reported the synthesis of nanorods using template-free infrared heating, a bottom-up fabrication approach with different powers such as 35%, 40%, 50%, 70%, and 100% with dicyanamide as precursor. The infrared heating greatly affects the morphology of the above synthesized GCN nanorods and also the heat provided by the infrared method is more effective than the heat by calcination method. Fifty milligrams of the prepared photocatalyst was added to 20 vol% of methanol that was used as sacrificial agent under a visible light (300-W Xe arc lamp) and the hydrogen evolution rate found to be 57.8 μmol/h. This was the first study for the synthesis of GCN nanorods by Templateless approach. Improvement in the charge-carrier separation and the elimination of the structural defects influences the intensity of the photocatalytic mechanism.

Zeng *et al.* (2016) reported the synthesis of 3D-network like morphology of GCN nanobelt by the hydrothermal treatment of bulk GCN at 160°C for 24 h. Twenty milligrams of the NPs added to 10 vol% triethanolamine as sacrificial agent that undergoes photocatalysis under visible light (300-W Xe arc clamp), the rate evolution hydrogen was reported to be 1.36 mmol/g/h. This was an eco-friendly and an easy method for the synthesis of GCN

Table 1.1 Summary of the GCN nanoparticles for dye degradation.

GCN synthesis method, size, and shape	Concentration of the catalyst bandgap	Dye, concentration	% Dye degradation	Light source, intensity (W)	References
GCN nanosheets; chemical exfoliation	25 mg, 2.92 eV	Methylene Blue (MB), 50,000 mg	94	Visible light (Xenon lamp), 500 W	[51]
GCN nanorods; template less infrared heating	10 mg, 2.5 eV	Hydrogen evolution	Rate of hydrogen evolution: 57.8 µmol/h	Visible light (Xe arc lamp), 300 W	[52]
GCN nanobelt; hydrothermal treatment of bulk GCN	25 mg, 2.84 eV	Hydrogen evolution	Hydrogen evolution rate: 1.36 mmol/g/h	Visible light (Xenon arc lamp), 300 W	[53]
Mesoporous GCN; calcination	40 mg, 2.69 eV	Rhodamine B (RhB), 5	100	Visible light (Halogen lamp), 300 W	[54]
2D GCN nanosheets; template growth	1.6 mg, 4.12 eV	MB, 5 mg	60	UV and visible light, 24 W	[55]
GCN 2D nanosheets; self-assembly from gaseous molecules	10 mg, 2.72 eV	MB, 30 mg	60	Visible light	[56]
GCN powder; thermal polymerization	300 mg, 2.75 eV at 550°C, 2.66 eV at 600°C, 2.71 eV at 650°C	RhB, 300,000	95	Sunlight	[57]
GCN nanosheets; calcination	50 mg, 2.78 eV at 500°C, 2.75 eV at 550°C, 2.82 eV at 600°C, 2.89 eV at 650°C	RhB, 10	18.34 (500°C) 22.20 (550°C) 94.83 (600°C) 94.80 (650°C)	Visible light (LED lamp), 50 W	[58]

nanobelt with higher redox capacity, decreased mass transfer resistance, and enhanced visible-light light driven photocatalytic activity.

Boorboor Azimi *et al.* (2018) used template-free method for the synthesis of mesoporous GCN, where melamine powder was placed in a muffle furnace at 550°C, and they used 40 mg of the NPs that is used to mineralization of RhB under visible light (300-W halogen lamp), and 100% of the dye was degraded below 60 min. This method yielded porous GCN with higher surface area, thus increasing the presence of active sites, which directly influences the increase in the photocatalytic activity.

Tiwari and Ram (2019) reported the synthesis of 2D nanosheets by template growth using Aloe Vera gel as a precursor for the biosynthetic route. The toxicity caused by the chemically synthesized GCN is reduced in this method of synthesis as these plants naturally act as a source of carbon and nitrogen. The prepared photocatalyst of 1.6 mg was utilized for the degradation of MB under both UV light (24-W light source) and visible light. Degradation efficiency of the dye was found to be minimum of 100% under UV light and 60% under visible light at 60 min.

Wen *et al.* (2019) claimed that the synthesis of 2D nanosheets through assembly of gaseous molecules using dicyanamide as a precursor. The synthesized GCN had more of nitrogen defects that improve the photocatalytic activity of the synthesized GCN. "Ten milligrams of the prepared photocatalyst was used to degrade MB under visible light (LED lights)," and 60% of the dye was degraded at the end of 60 min. The factors that enhance the photocatalytic activity of a 2D-structured, thin-layered GCN are higher separation efficiency of the charge carriers, altered band structure, and larger surface area.

Yadav *et al.* (2021) claimed the fabrication of GCN nano powder using thermal polymerization using melamine as precursor. The nano powder was synthesized "at different temperatures ranging from 550°C to 650°C, at an interval of 50°C." The various temperature ranges directly affect the morphological, optical, and structural properties. "Three hundred milligrams of the GCN was used for the degradation of RhB under sunlight, and 95% of the dye was degraded at 320 min."

Sewnet *et al.* (2023) reported a comparative study by synthesizing 2D nanosheets by calcination method using urea and thiourea as precursors. Nanosheets were prepared with different calcination temperatures ranging from 500°C to 650°C. These different temperatures for calcination affect the structure's crystallinity, surface area, and range of visible light absorption, which, in turn, improves the photocatalytic performance. Fifty milligrams of the 2D CGN was used for the degradation of RhB under visible light (50-W LED lamp), and 94.83% of the dye was degraded after 180 min of irradiation by the photocatalyst prepared at 600°C. In general, the calcination temperature affects the photocatalytic efficacy.

The morphology structure affected the photocatalytic efficacy and the stability of the photocatalyst that increased their photocatalytic activity. These studies showed that the modification in the process of synthesis, the precursors, and the temperatures affected the surface morphology by increasing the photocatalytic activity. Of the above-mentioned studies, mesoporous GCN showed higher efficiency in the dye degradation which is almost of 100% efficiency.

The powdered form of GCN is generally used that makes it unsuitable for its repeated usage [59]. Although GCN has many advantages over the metallic photocatalysts, they also have drawbacks such as higher recombination rates due to the presence of defects on its surface [60], reduced surface area, and lower absorption coefficient [61]. To make GCN an efficient photocatalysts, modifications are done such as doping [62], nanostructured engineering [59, 63], heterojunction construction [61], and bandgap engineering [64, 65].

1.3 GCN Modification to Act as a Photocatalyst

Doping is the addition of impurities to the semiconductors to improve their properties, and this is a common method used for modulating the bandgap of GCN [66]. The doping elements may be metal, non-metal, or halogen. Combining GCN with other highly conductive elements forming a heterojunction structure is helpful in the modification, which causes a combined effect of both the structures and enhances the activity. By rematching the energy band structure of two semiconductors, heterojunctions may increase the separation efficiency and alter charge transfer routes [16]. The heterojunctions formed maybe type I heterojunction, type II heterojunction, or Z-scheme heterojunctions. Nanostructured engineering including 0D quantum dots [67], 1D nanotube or nanorods [52], 2D nanosheets [68], and 3D structures such as mesoporous, microspheres, and nanobelts [67] is made to enhance the degradation activity. The nanostructure modification regulates the diffusion distance of holes and electrons, reaching the redox sites that become the reason for their enhanced photocatalytic activity [67]. Figure 1.4 presents the GCN structure with doping sites where the doping elements are added, thus enhancing its properties for dye degradation. Table 1.2 shows a few studies conducted for the dye degradation and hydrogen evolution using the doped GCN. Defect engineering can also enhance the photocatalytic activity, which creates higher catalytic reaction

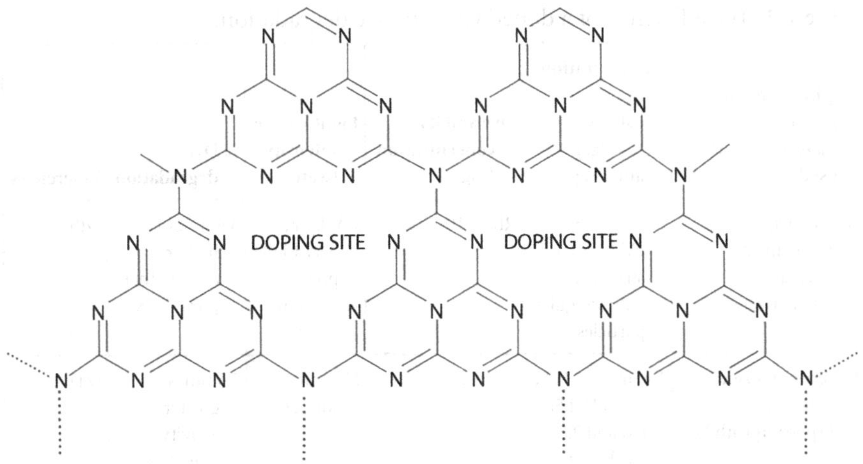

Figure 1.4 Doped GCN showing doping site.

on surface, efficient electron–hole separation, and enlarged light absorption of the photocatalyst [69].

Zhang *et al.* (2014) reported the use of an alkali metal, sodium (Na) as doping element to enhance the photocatalytic activity of GCN with NaOH and dicyanamide powder as precursors that are polymerized to form the photocatalyst. Interstitial doping is done rather than substitutional doping, as they cannot be doped in the ion state due to the larger size of the ion. Fifty milligrams of the photocatalyst was used for the degradation of RhB under UV light (250-W high-pressure sodium lamp). The doping of Na into GCN improved its photocatalytic activity by 3.5 times than GCN. Doping of sodium caused the synthesized GCN to have larger surface area, smaller grain size, greater separation rates of photogenerated electrons and holes, and narrow band energy gap. Also, the positioning of the valence band and conduction band can be done by regulating the concentration of the sodium doped into GCN.

Tonda *et al.* (2014) reported the use of iron (Fe) as doping element with ferric chloride and melamine as precursors. The oxidation state of the doping element is +3. The optical and electronic properties of the GCN nanosheets are greatly affected by Fe. Different concentrations of Fe were doped into the nanosheets prepared by liquid exfoliation that is used for a comparative study. One hundred milligrams of the photocatalyst was used for the degradation of RhB under sunlight and the results showed that the degradation was enhanced seven times greater than bulk GCN and 4.5 times greater than GCN nanosheets. The bulk GCN doped with Fe exhibits

Table 1.2 The efficiency of a doped GCN in dye degradation.

Doping element; precursor materials; method used	Concentration of the catalyst; bandgap size and shape	Dye and its concentration (mg)	Light source intensity (watt)	% Dye degradation	References
Na; NaOH, dicyanamide; thermal polymerization	50 mg; 2.58 eV; layered structure with irregular particles	Rh-B, 10	UV light; 250-W high pressure sodium lamp	3.5 times higher activity than pure GCN	[70]
Fe; ferric chloride, melamine; dispersion with Fe	100 mg; 2.64 eV, 2.61 eV, 2.56 eV, and 2.54 eV; Fe^{3+} in the nanosheet; 50 nm	Rh-B, 5	Natural sunlight	7.5 times greater activity than bulk GCN and four times better than pure GCN nanosheets	[71]
S; trithiocyanuric acid, melamine; direct thermal condensation	20 mg; 2.56 eV; microrods; 5 μ to 10 μm	H_2 evolution	Visible light; 500-W Xe lamp	5,000 μmol/g/h (hydrogen evolution rate)	[72]
N; hydrazine hydrate, melamine	80 mg; 2.65 eV; Thin layered structure	H_2 evolution	Visible Light; 300W Xe lamp	7.86μmol/h (hydrogen evolution rate)	[73]
Cu; $CuCl_2$, melamine; solution mixing and sonication	50 mg; 2.51 eV, 2.43 eV, 2.39 eV, 2.33 eV, 2.12 eV, and 2.25 eV; platelet-like structure; 1 μm	MO, 10	Visible light; 350-W Tungsten lamp	90.2%, 89.6%, 80.3%, 71.4%, 68.7%, 67.1%	[74]
Co; CoPc, melamine; thermal polymerization	50 mg; 2.66 eV and 2.62 eV; tacked lamellar morphology; 1 μm	H_2 evolution	Visible light; 300-W Xe lamp	28 μmol/h (hydrogen evolution rate)	[75]

lower photocatalytic activity compared to the Fe doped GCN nanosheets. The doped GCN obtained in this work had higher light absorption capacity and was stable even after repeated use.

Feng *et al.* (2014) in their study used sulfur (S), a non-metal as a doping element in the nano porous GCN. Synthesis of the sulfur-doped GCN was done using melamine and trithiocyanuric acid as precursors that are yellow in color. Melamine was used as a precursor because they can easily produce the product with high quality, and it has been reported in various works to be an efficient precursor for the production of GCN. Twenty milligrams of the photocatalyst was used for hydrogen evolution under visible light (500-W Xe lamp) with aqueous triethanolamine solution as sacrificial agent, and hydrogen evolution rate was 5,000 μmol/g/h. The visible-light absorption was increased, and, thus, its photocatalytic activity was also enhanced.

Fang *et al.* (2015), reported synthesis of GCN doped with nitrogen as doping element using hydrazine hydrate and melamine as precursors. Melamine here is pre-treated with hydrazine hydrate to produce more of nitrogen from it and, further, is calcined for the synthesis of nitrogen-doped GCN. Self-doping was done in this study that avoided the drawback of anion doping that is the formation of recombination center. Eighty milligrams of the photocatalyst was used for the hydrogen evolution in triethanolamine solution under visible light (300-W Xe lamp), and the rate of hydrogen evolution was found to be 7.86 μmol/h.

Le *et al.* (2016) reported the use of Copper (Cu) as a doping element using $CuCl_2$ and melamine as precursors, by polymerization followed by sonication. Nanoporous structures are formed that improves the visible light utilization, along with higher generation, separation, and transfer of the photogenerated carriers. Fifty milligrams of the photocatalyst was used for the degradation of methyl orange (MO) under visible light (300-W Xe lamp). Different concentrations of the doped Cu were prepared, and degradation was found to be 90.2%, 89.6%, 80.3%, 71.4%, 68.7%, 67.1%, 51.8%, and 19.7% for the different concentrations.

Chen *et al.* (2017) reported the use of cobalt (Co) as doping element. Cobalt was doped into the framework of GCN with cobalt phthalocyanine (CoPc) and melamine as precursors. The synthesis was followed by one-step thermal polycondensation process. Fifty milligrams of the prepared photocatalyst was added to triethanolamine solution for the hydrogen evolution under visible light (300-W Xe lamp), and the rate of hydrogen evolution was found to be 28 μmol/h. The electron distribution system is modified with the doping of cobalt in the framework of GCN, decreases the bandgap, and enhances the photocatalytic performance of the GCN.

Few of the studies done in the field of doped GCN photocatalyst for photocatalytic activity showed that they have an enhanced photocatalytic activity. This was because the doped element supresses the crystalline growth of GCN, which hinders its photocatalytic activity by the increase in its surface area. Thus, it was observed that doping can help in the enhancement of the photocatalytic activity. Fe-doped and Cu-doped GCN showed better efficiency in the above-mentioned studies.

1.4 GCN-Based Nanocomposites as a Photocatalyst

GCN nanocomposites have gained attention in the recent times as they have greater advantages when compared to GCN nanoparticles as single units. Nanocomposites are basically a complex structure consisting of two or more nanostructures combined, thus enhancing the properties of GCN. The combined effects of the prepared photocatalyst by nanocomposite have greater surface area with reduced size and enhanced photocatalytic effect. The light absorption capacity of the photocatalyst is also increased with the nanocomposite structure. Every material used as photocatalyst possesses its own drawbacks; thus, the nanocomposite structures help overcome these drawbacks by nullifying their drawbacks with the properties of the other materials in the composite structure. Thus, nanocomposite structures are studied for dye degradation. Figure 1.5 presents the general mechanism of dye degradation using GCN-based nanocomposite [76]. Table 1.3 shows the different GCN-based nanocomposites with their synthesis method and the dye degradation efficiency.

Xue *et al.* (2015) reported the Au/pg-C_3N_4/GR–layered composite synthesized using sonochemical method. It is a heterostructure plasmonic structure prepared using urea as precursor. This type of plasmonic photocatalysis has gained great importance these days due to their increased absorption of large range of UV and visible light. One hundred milligrams of the photocatalyst was utilized for the degradation of MB under visible Light (500-W Xe lamp); almost 100% of the dye was degraded after 180 min. The reaction sites were increased in this composite along with effective light harvesting, photogenerated electron separation and transportation. These factors together make this plasmonic structure an efficient photocatalyst.

Kumar *et al.* (2018) reported that g-C_3N_4/$CaTiO_3$ was fabricated followed by the modification using heat. It is a type of organic–inorganic heterojunction photocatalyst having flake-like morphology. Twenty-five milligrams of the photocatalyst was used for the degradation of RhB under

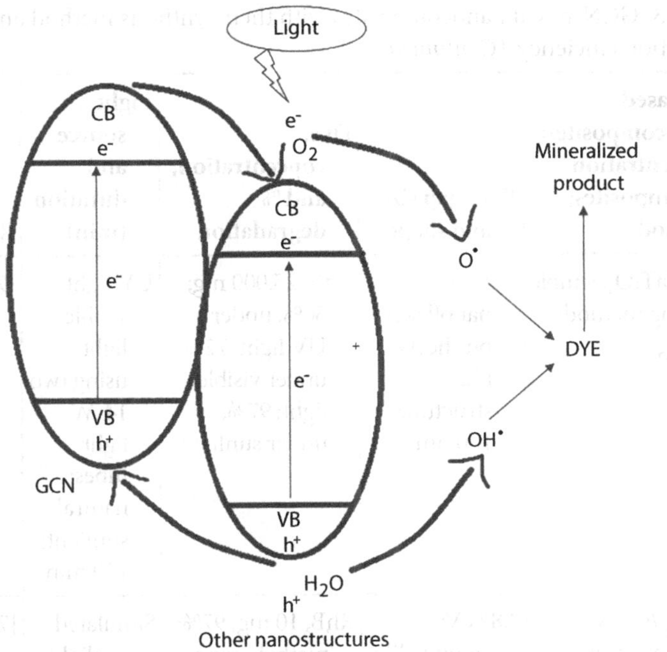

Figure 1.5 Schematic representation of dye degradation using GCN-based nanocomposites.

Table 1.3 GCN-based nanocomposites with their synthesis method and the dye degradation efficiency.

GCN-based nanocomposites; concentration of composites; method	Bandgap size and shape	Dye, concentration, and % degradation	Light source and duration (min)	References
Au/p-GCN/GR; sonication and photo deposition; 100 mg	2.7 eV; spherical; 5 nm	MB, 10 mg, 100%	Visible light; 500-W Xenon lamp; 180 min	[77]

(*Continued*)

Table 1.3 GCN-based nanocomposites with their synthesis method and the dye degradation efficiency. (*Continued*)

GCN-based nanocomposites; concentration of composites; method	Bandgap size and shape	Dye, concentration, and % degradation	Light source and duration (min)	References
GCN/$CaTiO_3$; simple mixing method; 25 mg	2.79–3.22 eV; nanoflakes on sheets like structure; 100 nm	RhB, 25,000 mg; 50%, under UV light; 72%, under visible light; 97%, under sunlight	UV light; visible light using two 14-W light tubes; natural sunlight; 120 min	[78]
$Bi_{12}TiO_{20}$/GCN; calcination and ultrasound; 200 mg	2.82 eV; granum-like structure; 10 nm	RhB, 10 mg, 97%; methyl orange, 20 mg, 90%	Simulated sunlight source; 500-W Xenon lamp;50 min	[79]
GCN/ZnO/ Mg-Al LDH; hydrothermal method; 10 mg	2.82 eV; ring-shaped patterns with hexagonally arranged brighter spots; 50 nm	MB, 96.5%	Visible light; 115 min	[80]

(*Continued*)

Table 1.3 GCN-based nanocomposites with their synthesis method and the dye degradation efficiency. (*Continued*)

GCN-based nanocomposites; concentration of composites; method	Bandgap size and shape	Dye, concentration, and % degradation	Light source and duration (min)	References
GCN/MoO_3; wet chemical method; 50 mg	4.5eV; ultrathin GCN nanosheets covering the MoO_3 nanobelt structure; 200 nm	RhB, 20 mg, 93%	Visible light; 150-W Xe lamp; 180 min	[81]
Phosphorous(P) and sulfur (S) co-doped with GCN PSCN/Ag@AgI/WO_3; one-pot precipitation method; 50 mg	PSCN, 2.51; AgI, 2.78; WO_3, 2.68; cloudy structure upon sheets of PSCN; 5 nm	Malachite green (MG), 1 mg, 90%	Visible light; 35-W LED lamp; 60 min	[82]
GCN@Ag_2ZrO_3; co-precipitation; 25 mg	2.4 eV; 106 nm; irregular spherical shaped Ag_2ZrO_3 decorated on wrinkled GCN nanosheets	RhB, 10 mg, 99.70%	Visible light; 100-W tungsten bulb; 120 min	[83]

(*Continued*)

Table 1.3 GCN-based nanocomposites with their synthesis method and the dye degradation efficiency. (*Continued*)

GCN-based nanocomposites; concentration of composites; method	Bandgap size and shape	Dye, concentration, and % degradation	Light source and duration (min)	References
GCN–CuO/ZnO; solvent mixing assisted with sonication; 50 mg	2.2eV; micron-size flower-like structure of metal oxide decorated on GCN	MB, 25,000, 99%	Visible light; 250-mV LED lamp; 75 min	[84]
Ce^{3+}-GCN adsorption method; 0.12 mg	3.92 eV; Ce^{3+} clusters on GCN nanosheets	MB, 0.015, 56% visible light, 95% sunlight	Sunlight; 180 min visible light; 200-W mercury lamp; 80 min	[85]
WMO@C/GCN; one-step hydrothermal scheme; 50 mg	2.43 eV; nanowires; 3.53 nm	MB, 5, 91%; MG, 5. 92%; Crystal Violet (CV), 5, 89%	Visible light (200-W bulb with tungsten filament); 120 min	[86]

sunlight, UV light (12 UV lamps), and visible light (two 14-W LED light), and 97% was degraded in sunlight, 50% in UV light, and 72% in visible light at 120 min. The heterojunction formation improves the bandgap that greatly affects the photocatalytic activity of the synthesized composite.

Cao *et al.* (2019) reported the $Bi_{12}TiO_{20}/g\text{-}C_3N_4$ produced using ultrasonic solution dispersion method, and the produced composite had a granum-like structure. Two hundred milligrams of the composite photocatalyst was used for the degradation of RhB and MO under simulated sunlight (500-W Xe lamp), and 97% of RhB and 90% of MO were degraded at the end of 50 min. Charge transformation efficiency was enhanced, and also its sustainability and reusability were amplified.

Bhuvaneswari *et al.* (2019) reported g-C_3N_4/ZnO/Mg-Al layered double hydroxide (LDH) by simple hydrothermal method that is an intercalated structure. In this structure, ZnO particles were fixed in the interlayer surface of LDH, and ZnO/Mg-Al LDHs are spread over the surface of g-C_3N_4. Ten milligrams of the photocatalyst was utilized for the degradation of MB under visible light, and 96.5% of the dye was degraded at the end of 115 min. The synthesized nanocomposite was a nitrogen-rich material that increases the reactive sites in the composite and also reducing the charge recombination, thus enhancing the photocatalytic activity of the nanocomposite.

Adhikari and Kim (2020) reported that the C_3N_4/MoO_3 nanocomposite, MoO_3, is considered to have great chemical and that thermal stability was synthesized by hydrothermal method, while GCN was prepared by direct calcination, and the composite was prepared by wet chemical method. Fifty milligrams of the prepared photocatalyst was used for the degradation of RhB under visible light (150-W Xe lamp), and 93% of the dye was degraded at 180 min. The charge carrier separation between C_3N_4 and MoO_3 is enhanced, which increase the adsorption capacity of the composite, thus increasing its photocatalytic activity.

Hasija *et al.* (2020) reported the PSCN/Ag@AgI/WO_3 that consists of phosphorous (P) and sulfur (S) co-doped with GCN coupled with Ag@AgI/WO_3, which was prepared in a single pot. Fifty milligrams of the photocatalyst was used for the degradation of MG under visible light (35-W LED lamp), and 90% of the dye was degraded after 60 min of irradiation. The structure was found to separate the electron–hole pair along with enlarged absorption in the visible light spectra, hence increasing the photocatalytic activity. The so-obtained photocatalyst had a dual Z-scheme mechanism for the photocatalytic activity.

Das *et al.* (2020) reported g-C_3N_4@Ag_2ZrO_3 with Ag_2ZrO_3 on the surface of the graphitic sheets by co-precipitation method. The study was conducted with different concentration of GCN in the composite, and all the prepared photocatalyst was used to test the degradation activity. Twenty-five milligrams of the photocatalyst was used for the degradation of RhB under visible light (100-W tungsten lamp), and 90.7% of the dye was degraded at 120 min. The GCN over nano Ag_2ZrO_3 structure improves the photocatalytic activity of the synthesized photocatalyst by making it more efficient in visible light harvesting.

Sivasakthi and Gurunathan (2020) reported g-C_3N_4-CuO/ZnO nanocomposite by following simple mixing, and the powder form of the nanocomposite was obtained with micron-size flower-like structure that enhances its properties. Fifty milligrams of the photocatalyst was used for the degradation of MB under visible light (250-mV LED lamp), and 99%

degradation of dye was observed after 75 min. The nanopetal flower-like structure enhances the light harvesting efficiency and also the charge separation efficiency, thus having greater photocatalytic activity of the prepared photocatalyst.

Kuila *et al.* (2020) reported Ce^{3+}-g-C_3N_4 by simple adsorption method. The Ce^{3+} ions are absorbed onto the 2D-layered GCN through chemisorption. The bandgap was calculated to be 3.92 eV. The photocatalyst of 0.12 mg was used for the degradation of MB under sunlight (8 h) and visible light (200-W mercury lamp) for 180 min, and 95% and 56% of the dye were degraded at respective conditions. The photocatalyst prepared exhibited increased surface area, reduced photorecombination probability, and better charge transfer ability, thus enhancing its photocatalytic activity.

Shaheen *et al.* (2023) reported the synthesis of a ternary nanocomposite, WMO@C/GCN by one-step hydrothermal method. Fifty milligrams of the prepared photocatalyst was used for the degradation of MB, MG, and CV under visible light (200-W bulb with tungsten filament); 91% of MB, 92% of MG, and 89% of CV were degraded over 120 min. The increased photocatalytic activity was associated with slower electron–hole recombination, increased surface area, higher redox ability due to the stable bonds formed between the GCN and WMO@C nanowire, and efficient charge transfer efficiency.

GCN-based nanocomposites that are mentioned in the above studies helps in the better degradation of the target dyes that are hazardous to human life and environment. These nanocomposites enhanced the degradation efficiency, due to their combined effect on the properties of the nanocomposite structure. The commonly chosen dyes were MB, MG, RhB, and MO in the above studies as they are the most commonly used dyes in the industries. Most of these composites efficiently degraded the target dyes under sunlight, visible light, and UV lights as per the studies mentioned above. Au/p-GCN/GR nanocomposite showed higher degradation efficiency of 100% when compared to the other nanocomposites mentioned above.

1.5 Conclusion

Clean water is very important for the safety of environment and of the human race. There are many conventional methods for wastewater treatment, but they have their own disadvantages. Thus, AOPs has been developed as an efficient method for wastewater treatment. Photocatalysis is one of the most used AOPs because of its advantages such as low cost, easy synthesis, and efficiency. Photocatalysts are very important for the photocatalysis process, and, in this review, GCN was the focus of study. It is an excellent non-metal polymeric photocatalyst with many advantages, yet they have a need for modification to improve their photocatalytic properties. GCN doped with metal atoms or other elements increases its efficiency of degradation as the properties of the doping element also improve the overall properties of the prepared GCN photocatalysts. Nanoparticles are gaining importance these days because of their numerous advantages, and, when they are combined with GCN, they increase the catalytic activity of the composite structure. These composite structures are preferred and studied these days because of their higher efficiency, less toxicity, easy synthesis, and other advantages. These modified photocatalysts are used for the dye degradation as discussed in the above content. This chapter gives an outline about the GCN and their application in the dye degradation in wastewater.

1.6 Future Perspectives

The area of photocatalysts is gaining more research interest; especially GCN has gained importance for research due to its advantages and efficient photocatalysis process. Modification for better photocatalytic activity is another area where a lot of composite structures can be built, and the methods of modification can also be improved to limit their disadvantages. The solar energy utilization is done by the GCN as a photocatalyst, but this can still be improved, thus increasing its absorption efficiency. The application of GCN and GCN-based nanocomposites can be studied still more, thus expanding the horizons for the application of GCN-based materials. Although there are numerous photocatalysts studied and used, there are many other structures that can have much more advantages than the present engineered materials. GCN used as photocatalyst is a wide area of research that has to be explored as an efficient photocatalytic material.

References

1. Kalikeri, S. and Shetty Kodialbail, V., Auto-combustion synthesis of narrow band-gap bismuth ferrite nanoparticles for solar photocatalysis to remediate azo dye containing water. *Environ. Sci. Pollut. Res.*, 28, 10, 12144–12152, 2021, doi: 10.1007/s11356-020-10879-w.

2. Rajasri, S. *et al.*, Enhanced photodegradation of 2,4-dinitrophenol by n–p type TiO2/BiOI nanocomposite. *J. Indian Chem. Soc.*, 99, 3, 1, 2022, doi: 10.1016/j.jics.2021.100337.
3. Kalikeri, S. and Kodialbail, V.S., Visible light active Bismuth ferrite embedded TiO2 nanocomposite structures for dye mineralization by photocatalysis -A strategy to harness solar energy for remediation of water contaminated with mixture of dyes. *Surf. Interfaces*, 36, 102492, 2023, doi: https://doi.org/10.1016/j.surfin.2022.102492.
4. Kalikeri, S. and Shetty Kodialbail, V., Solar light-driven photocatalysis using mixed-phase bismuth ferrite (BiFeO3/Bi25FeO40) nanoparticles for remediation of dye-contaminated water: kinetics and comparison with artificial UV and visible light-mediated photocatalysis. *Environ. Sci. Pollut. Res.*, 25, 14, 13881–13893, 2018, doi: 10.1007/s11356-018-1291-0.
5. Kalikeri, S., Kamath, N., Gadgil, D.J., Shetty Kodialbail, V., Visible light-induced photocatalytic degradation of Reactive Blue-19 over highly efficient polyaniline-TiO2 nanocomposite: a comparative study with solar and UV photocatalysis. *Environ. Sci. Pollut. Res.*, 25, 4, 3731–3744, 2018, doi: 10.1007/s11356-017-0663-1.
6. Hube, S. *et al.*, Direct membrane filtration for wastewater treatment and resource recovery: A review. *Sci. Total Environ.*, 710, 3, 2020, doi: 10.1016/j.scitotenv.2019.136375.
7. Ahmad, N.A., Goh, P.S., Karim, Z.A., Ismail, A.F., Thin film composite membrane for oily waste water treatment: Recent advances and challenges. *Membranes (Basel)*, 8, 4, 2, 2018, doi: 10.3390/membranes8040086.
8. Luo, J. *et al.*, Potential influences of exogenous pollutants occurred in waste activated sludge on anaerobic digestion: A review. *J. Hazard. Mater.*, 383, 1, 2020, doi: 10.1016/j.jhazmat.2019.121176.
9. Hadi, P., Xu, M., Ning, C., Lin, C.S.K., McKay, G., A critical review on preparation, characterization and utilization of sludge-derived activated carbons for wastewater treatment. *Chem. Eng. J.*, 260, 1, 2015, doi: 10.1016/j.cej.2014.08.088.
10. Riaz, S. and Park, S.J., An overview of TiO2-based photocatalytic membrane reactors for water and wastewater treatments. *J. Ind. Eng. Chem.*, 84, 35, 2020, doi: 10.1016/j.jiec.2019.12.021.
11. Kumari, P., Bahadur, N., Dumée, L.F., Photo-catalytic membrane reactors for the remediation of persistent organic pollutants – A review. *Sep. Purif. Technol.*, 230, 2-4, 2020, doi: 10.1016/j.seppur.2019.115878.
12. Chen, X., Huang, G., An, C., Feng, R., Wu, Y., Huang, C., Superwetting polyethersulfone membrane functionalized with ZrO2 nanoparticles for polycyclic aromatic hydrocarbon removal. *J. Mater. Sci. Technol.*, 98, 22, 2022, doi: 10.1016/j.jmst.2021.01.063.
13. Qi, K., Liu, S.Y., Zada, A., Graphitic carbon nitride, a polymer photocatalyst. *J. Taiwan Inst. Chem. Eng.*, 109, 111–123, 2020, doi: 10.1016/j.jtice.2020.02.012.

14. Dewil, R., Mantzavinos, D., Poulios, I., Rodrigo, M.A., New perspectives for Advanced Oxidation Processes. *J. Environ. Manag.*, 195, 1, 2017, doi: 10.1016/j.jenvman.2017.04.010.
15. Paździor, K., Bilińska, L., Ledakowicz, S., A review of the existing and emerging technologies in the combination of AOPs and biological processes in industrial textile wastewater treatment. *Chem. Eng. J.*, 376, 9, 10, 2019, doi: 10.1016/j.cej.2018.12.057.
16. Guo, X., Duan, J., Wang, W., Zhang, Z., Modified graphitic carbon nitride as the photocatalyst for wastewater treatment under visible light irradiation. *Fuel*, 280, March, 118544, 2020, doi: 10.1016/j.fuel.2020.118544.
17. Kabra, K., Chaudhary, R., Sawhney, R.L., Treatment of hazardous organic and inorganic compounds through aqueous-phase photocatalysis: A review. *Ind. Eng. Chem. Res.*, 43, 24, 7683, 2004, doi: 10.1021/ie0498551.
18. Ismael, M., A review on graphitic carbon nitride (g-C3N4) based nanocomposites: Synthesis, categories, and their application in photocatalysis. *J. Alloys Compd.*, 846, 156446, 2020, doi: 10.1016/j.jallcom.2020.156446.
19. Niu, P., Dai, J., Zhi, X., Xia, Z., Wang, S., Li, L., Photocatalytic overall water splitting by graphitic carbon nitride. *InfoMat*, 3, 9, 931–961, 2021, doi: 10.1002/inf2.12219.
20. Abe, R., (Invited) Photocatalytic and Photoelectrochemical Water Splitting Under Visible Light Irradiation. *ECS Meet. Abstr.*, MA2016-02, 49, 1, 2016, doi: 10.1149/ma2016-02/49/3607.
21. Rather, R.A., Mehta, A., Lu, Y., Valant, M., Fang, M., Liu, W., Influence of exposed facets, morphology and hetero-interfaces of BiVO4 on photocatalytic water oxidation: A review. *Int. J. Hydrogen Energy*, 46, 42, 3, 2021, doi: 10.1016/j.ijhydene.2021.04.060.
22. Liu, C., Xiong, M., Chai, B., Yan, J., Fan, G., Song, G., Construction of 2D/2D Ni2P/CdS heterojunctions with significantly enhanced photocatalytic H2 evolution performance. *Catal. Sci. Technol.*, 9, 24, 6929–6937, 2019, doi: 10.1039/C9CY02045D.
23. Li, J.Y., Li, Y.H., Zhang, F., Tang, Z.R., Xu, Y.J., Visible-light-driven integrated organic synthesis and hydrogen evolution over 1D/2D CdS-Ti3C2Tx MXene composites. *Appl. Catal B*, 269, 1, 2020, doi: 10.1016/j.apcatb.2020.118783.
24. Hara, M., Hitoki, G., Takata, T., Kondo, J.N., Kobayashi, H., Domen, K., TaON and Ta3N5 as new visible light driven photocatalysts. *Catal. Today*, 78, 556, 2003, doi: 10.1016/S0920-5861(02)00354-1.
25. Hitoki, G., Ishikawa, A., Takata, T., Kondo, J.N., Hara, M., Domen, K., ChemInform Abstract: Ta3N5 as a Novel Visible Light-Driven Photocatalyst ($\lambda < 600$ nm). *Chem. Inform.*, 33, 736, 45, 2010, doi: 10.1002/chin.200245007.
26. Chumha, N. *et al.*, Photocatalytic activity of CuInS2 nanoparticles synthesized via a simple and rapid microwave heating process. *Mater. Res. Express*, 7, 1, 3, 2020, doi: 10.1088/2053-1591/ab6885.
27. Hu, J., Chen, C., Zheng, Y., Zhang, G., Guo, C., Li, C.M., Spatially Separating Redox Centers on Z-Scheme ZnIn2S4/BiVO4 Hierarchical Heterostructure

for Highly Efficient Photocatalytic Hydrogen Evolution. *Small*, 16, 37, 1, 2020, doi: 10.1002/smll.202002988.
28. Lee, G.J., Zheng, Y.C., Wu, J.J., Fabrication of hierarchical bismuth oxyhalides (BiOX, X = Cl, Br, I) materials and application of photocatalytic hydrogen production from water splitting. *Catal. Today*, 307, 1, 2018, doi: 10.1016/j.cattod.2017.04.044.
29. Cui, C. *et al.*, Fast assembly of Ag3PO4 nanoparticles within three-dimensional graphene aerogels for efficient photocatalytic oxygen evolution from water splitting under visible light. *Appl. Catal. B*, 200, 667, 2017, doi: 10.1016/j.apcatb.2016.07.056.
30. Kibria, M.G. and Mi, Z., Artificial photosynthesis using metal/nonmetal-nitride semiconductors: Current status, prospects, and challenges. *J. Mater. Chem. A*, 4, 8, 3, 2016, doi: 10.1039/c5ta07364b.
31. Ochiai, T. and Fujishima, A., Photoelectrochemical properties of TiO2 photocatalyst and its applications for environmental purification. *J. Photochem. Photobiol. C*, 13, 4, 248, 2012, doi: 10.1016/j.jphotochemrev.2012.07.001.
32. Aslam, I. *et al.*, The synergistic effect between WO3 and g-C3N4 towards efficient visible-light-driven photocatalytic performance. *New J. Chem.*, 38, 11, 5462–5469, 2014, doi: 10.1039/C4NJ01370K.
33. Guo, J. *et al.*, Experimental and DFT insights of the Zn-doping effects on the visible-light photocatalytic water splitting and dye decomposition over Zn-doped BiOBr photocatalysts. *Appl. Catal. B*, 243, 503, 2019, doi: 10.1016/j.apcatb.2018.09.089.
34. Abe, R., Recent progress on photocatalytic and photoelectrochemical water splitting under visible light irradiation. *J. Photochem. Photobiol. C*, 11, 4, 181, 2010, doi: 10.1016/j.jphotochemrev.2011.02.003.
35. Kudo, A. and Miseki, Y., Heterogeneous photocatalyst materials for water splitting. *Chem. Soc Rev.*, 38, 1, 255, 2009, doi: 10.1039/b800489g.
36. Zhao, G., Yang, H., Liu, M., Xu, X., Metal-free graphitic carbon nitride photocatalyst goes into two-dimensional time. *Front. Chem.*, 6, DEC, 1–11, 2018, doi: 10.3389/fchem.2018.00551.
37. Xu, Y., Kraft, M., Xu, R., Metal-free carbonaceous electrocatalysts and photocatalysts for water splitting. *Chem. Soc Rev.*, 45, 11, 3040, 2016, doi: 10.1039/c5cs00729a.
38. Naseri, A., Samadi, M., Pourjavadi, A., Moshfegh, A.Z., Ramakrishna, S., Graphitic carbon nitride (g-C 3 N 4)-based photocatalysts for solar hydrogen generation: recent advances and future development directions. *J. Mater. Chem. A Mater.*, 5, 45, 23406–23433, 2017, doi: 10.1039/C7TA05131J.
39. Bandyopadhyay, A., Ghosh, D., Kaley, N.M., Pati, S.K., Photocatalytic Activity of g-C3N4 Quantum Dots in Visible Light: Effect of Physicochemical Modifications. *J. Phys. Chem. C*, 121, 3, 1982–1989, Jan. 2017, doi: 10.1021/acs.jpcc.6b11520.
40. Jun, Y.S., Park, J., Lee, S.U., Thomas, A., Hong, W.H., Stucky, G.D., Three-dimensional macroscopic assemblies of low-dimensional carbon nitrides for

enhanced hydrogen evolution. *Angew. Chem. Int. Ed.*, 52, 42, 11083, 2013, doi: 10.1002/anie.201304034.

41. Sun, J., Zhang, J., Zhang, M., Antonietti, M., Fu, X., Wang, X., Bioinspired hollow semiconductor nanospheres as photosynthetic nanoparticles. *Nat. Commun.*, 3, 3, 2012, doi: 10.1038/ncomms2152.
42. Bai, X., Wang, L., Zong, R., Zhu, Y., Photocatalytic Activity Enhanced via g-C3N4 Nanoplates to Nanorods. *J. Phys. Chem. C*, 117, 19, 9952–9961, May 2013, doi: 10.1021/jp402062d.
43. Wang, X. *et al.*, A metal-free polymeric photocatalyst for hydrogen production from water under visible light. *Nat. Mater.*, 8, 1, 76, 2009, doi: 10.1038/nmat2317.
44. Zhang, Y., Mori, T., Ye, J., Polymeric carbon nitrides: Semiconducting properties and emerging applications in photocatalysis and photoelectrochemical energy conversion. *Sci. Adv. Mater.*, 4, 2, 282, 2012, doi: 10.1166/sam.2012.1283.
45. Cao, S., Low, J., Yu, J., Jaroniec, M., Polymeric Photocatalysts Based on Graphitic Carbon Nitride. *Adv. Mater.*, 27, 13, 2164, 2015, doi: 10.1002/adma.201500033.
46. Huang, Z., Li, F., Chen, B., Lu, T., Yuan, Y., Yuan, G., Well-dispersed g-C3N4 nanophases in mesoporous silica channels and their catalytic activity for carbon dioxide activation and conversion. *Appl. Catal. B*, 136–137, 2013, doi: 10.1016/j.apcatb.2013.01.057.
47. Zhang, Y. *et al.*, Synthesis and luminescence mechanism of multicolor-emitting g-C3N4 nanopowders by low temperature thermal condensation of melamine. *Sci. Rep.*, 3, 1, 2013, doi: 10.1038/srep01943.
48. Wang, L., Wang, K., He, T., Zhao, Y., Song, H., Wang, H., Graphitic carbon nitride-based photocatalytic materials: Preparation strategy and application. *ACS Sustain. Chem. Eng.*, 8, 43, 16048–16085, 2020, doi: 10.1021/acssuschemeng.0c05246.
49. Qiao, S. *et al.*, Pore confinement effects and stabilization of carbon nitride oligomers in macroporous silica for photocatalytic hydrogen production. *Carbon N Y.*, 106, 321, 2016, doi: 10.1016/j.carbon.2016.05.039.
50. Inoki, H., Seo, G., Kanai, K., Synthesis of graphitic carbon nitride under low ammonia partial pressure. *Appl. Surf. Sci.*, 534, 3, 2020, doi: 10.1016/j.apsusc.2020.147569.
51. Xu, J., Zhang, L., Shi, R., Zhu, Y., Chemical exfoliation of graphitic carbon nitride for efficient heterogeneous photocatalysis. *J. Mater. Chem. A Mater.*, 1, 46, 3, 4, 2013, doi: 10.1039/c3ta13188b.
52. Li, H.-J., Qian, D.-J., Chen, M., Templateless Infrared Heating Process for Fabricating Carbon Nitride Nanorods with Efficient Photocatalytic H2 Evolution. *ACS Appl. Mater. Interfaces*, 7, 45, 25162–25170, Nov. 2015, doi: 10.1021/acsami.5b06627.
53. Zeng, Y. *et al.*, A three-dimensional graphitic carbon nitride belt network for enhanced visible light photocatalytic hydrogen evolution. *J. Mater. Chem. A Mater.*, 4, 48, 2, 3, 2016, doi: 10.1039/C6TA07397B.

54. Boorboor Azimi, E., Badiei, A., Hossaini Sadr, M., Amiri, A., A template-free method to synthesize porous G-C3N4 with efficient visible light photodegradation of organic pollutants in water. *Adv. Powder Technol.*, 29, 11, 2785, 2787, 2790, 2018, doi: 10.1016/j.apt.2018.07.027.
55. Tiwari, B. and Ram, S., Biogenic Synthesis of Graphitic Carbon Nitride for Photocatalytic Degradation of Organic Dyes. *ACS Omega*, 4, 6, 10264, 10269, 10270, 2019, doi: 10.1021/acsomega.9b00989.
56. Wen, Y. *et al.*, Graphitic carbon nitride nanosheets prepared by gaseous molecules assembling for enhanced photocatalytic performance. *J. Mater. Sci.*, 54, 2, 1463, 1464, 1470, 2019, doi: 10.1007/s10853-018-2937-5.
57. Yadav, A.A., Kang, S.W., Hunge, Y.M., Photocatalytic degradation of Rhodamine B using graphitic carbon nitride photocatalyst. *J. Mater. Sci.: Mater. Electron.*, 32, 11, 15577–15585, 2021, doi: 10.1007/s10854-021-06106-y.
58. Sewnet, A. *et al.*, Single-Step Synthesis of Graphitic Carbon Nitride Nanomaterials by Directly Calcining the Mixture of Urea and Thiourea: Application for Rhodamine B (RhB) Dye Degradation. *Nanomaterials*, 13, 4, 1, 4, 5, 10, 2023, doi: 10.3390/nano13040762.
59. Reddy, K.R., Reddy, C.V., Nadagouda, M.N., Shetti, N.P., Jaesool, S., Aminabhavi, T.M., Polymeric graphitic carbon nitride (g-C 3 N 4)-based semiconducting nanostructured materials: Synthesis methods, properties and photocatalytic applications. *J. Environ. Manag.*, 238, February, 25–40, 2019, doi: 10.1016/j.jenvman.2019.02.075.
60. Tu, W. *et al.*, Investigating the Role of Tunable Nitrogen Vacancies in Graphitic Carbon Nitride Nanosheets for Efficient Visible-Light-Driven H2 Evolution and CO2 Reduction. *ACS Sustain. Chem. Eng.*, 5, 8, 7260, 7261, 2017, doi: 10.1021/acssuschemeng.7b01477.
61. Li, H., Liu, J., Hou, W., Du, N., Zhang, R., Tao, X., Synthesis and characterization of g-C3N4/Bi2MoO6 heterojunctions with enhanced visible light photocatalytic activity. *Appl. Catal B*, 160–161, 1, 89, 2014, doi: 10.1016/j.apcatb.2014.05.019.
62. Gao, J., Wang, Y., Zhou, S., Lin, W., Kong, Y., A Facile One-Step Synthesis of Fe-Doped g-C3N4 Nanosheets and Their Improved Visible-Light Photocatalytic Performance. *ChemCatChem*, 9, 9, 1708, 2017, doi: 10.1002/cctc.201700492.
63. Zhu, Y.-P., Ren, T.-Z., Yuan, Z.-Y., Mesoporous Phosphorus-Doped g-C3N4 Nanostructured Flowers with Superior Photocatalytic Hydrogen Evolution Performance. *ACS Appl. Mater. Interfaces*, 7, 30, 16850–16856, Aug. 2015, doi: 10.1021/acsami.5b04947.
64. Ji, M., Di, J., Ge, Y., Xia, J., Li, H., 2D-2D stacking of graphene-like g-C3N4/Ultrathin Bi4O5Br2 with matched energy band structure towards antibiotic removal. *Appl. Surf. Sci.*, 413, 372–380, 2017, doi: https://doi.org/10.1016/j.apsusc.2017.03.287.
65. Liu, C., Zhang, Y., Dong, F., Du, X., Huang, H., Easily and Synchronously Ameliorating Charge Separation and Band Energy Level in Porous g-C3N4

for Boosting Photooxidation and Photoreduction Ability. *J. Phys. Chem. C*, 120, 19, 10381–10389, May 2016, doi: 10.1021/acs.jpcc.6b01705.
66. Phoon, B.L. *et al.*, Co-Doped, Tri-Doped, and Rare-Earth-Doped g-C3N4 for Photocatalytic Applications: State-of-the-Art. *Catalysts*, 12, 6, 5, 2022, doi: 10.3390/catal12060586.
67. Mishra, A., Mehta, A., Basu, S., Shetti, N.P., Reddy, K.R., Aminabhavi, T.M., Graphitic carbon nitride (g–C3N4)–based metal-free photocatalysts for water splitting: A review. *Carbon N Y.*, 149, 693–721, 2019, doi: 10.1016/j.carbon.2019.04.104.
68. Huang, D. *et al.*, Megamerger in photocatalytic field: 2D g-C3N4 nanosheets serve as support of 0D nanomaterials for improving photocatalytic performance. *Appl. Catal. B: Environ.*, 240, 153, 2019, doi: 10.1016/j.apcatb.2018.08.071.
69. Liao, J. *et al.*, Nitrogen defect structure and NO+ intermediate promoted photocatalytic NO removal on H2 treated g-C3N4. *Chem. Eng. J.*, 379, 2, 2020, doi: 10.1016/j.cej.2019.122282.
70. Zhang, J., Hu, S., Wang, Y., A convenient method to prepare a novel alkali metal sodium doped carbon nitride photocatalyst with a tunable band structure. *RSC Adv.*, 4, 108, 62912, 62913, 62919, 2014, doi: 10.1039/c4ra11377b.
71. Tonda, S., Kumar, S., Kandula, S., Shanker, V., Fe-doped and -mediated graphitic carbon nitride nanosheets for enhanced photocatalytic performance under natural sunlight. *J. Mater. Chem. A Mater.*, 2, 19, 6773, 6776, 6779, 2014, doi: 10.1039/c3ta15358d.
72. Feng, L.L. *et al.*, Nanoporous sulfur-doped graphitic carbon nitride microrods: A durable catalyst for visible-light-driven H2 evolution. *Int. J. Hydrogen Energy*, 39, 28, 15373–15375, 2014, doi: 10.1016/j.ijhydene.2014.07.160.
73. Fang, J., Fan, H., Li, M., Long, C., Nitrogen self-doped graphitic carbon nitride as efficient visible light photocatalyst for hydrogen evolution. *J. Mater. Chem. A Mater.*, 3, 26, 13820, 13825 2015, doi: 10.1039/c5ta02257f.
74. Le, S. *et al.*, Cu-doped mesoporous graphitic carbon nitride for enhanced visible-light driven photocatalysis. *RSC. Adv.*, 6, 45, 38813, 38815, 38816, 38818, 2016, doi: 10.1039/c6ra03982k.
75. Chen, P.W., Li, K., Yu, Y.X., De Zhang, W., Cobalt-doped graphitic carbon nitride photocatalysts with high activity for hydrogen evolution. *Appl. Surf. Sci.*, 392, 608, 609, 613, 614, 2017, doi: 10.1016/j.apsusc.2016.09.086.
76. Akintunde, O.O., Yu, L., Hu, J., Kibria, M.G., Achari, G., Visible-Light Driven Photocatalytic Degradation of 4-Chlorophenol Using Graphitic Carbon Nitride-Based Nanocomposites. *Catalysts*, 12, 3, 10, 2022, doi: 10.3390/catal12030281.
77. Xue, J., Ma, S., Zhou, Y., Wang, Q., Au-loaded porous graphitic C3N4/graphene layered composite as a ternary plasmonic photocatalyst and its visible-light photocatalytic performance. *RSC Adv.*, 5, 107, 88250, 88251, 88253, 88254, 88256, 2015, doi: 10.1039/c5ra17719g.

78. Kumar, A., Schuerings, C., Kumar, S., Kumar, A., Krishnan, V., Perovskite-structured CaTiO3 coupled with g-C3N4 as a heterojunction photocatalyst for organic pollutant degradation. *Beilstein J. Nanotechnol.*, 9, 1, 671, 673, 674, 676, 679, 682, 684, 2018, doi: 10.3762/bjnano.9.62.
79. Cao, L. *et al.*, A novel Bi12TiO20/g-C3N4 hybrid catalyst with a bionic granum configuration for enhanced photocatalytic degradation of organic pollutants. *J. Hazard Mater.*, 379, 1–3, 7, 9, 2019, doi: 10.1016/j.jhazmat.2019.120808.
80. Bhuvaneswari, K., Palanisamy, G., Pazhanivel, T., Maiyalagan, T., Bharathi, G., Photodegradation Activity of Nitrogen-rich Graphitic Carbon Nitride Intercalated ZnO\Mg-Al Layered Double Hydroxide Ternary Nanocomposites on Methylene Blue Dye. *ChemistrySelect*, 4, 11, 2982, 2986, 2987, 2988, 2019, doi: 10.1002/slct.201900146.
81. Adhikari, S. and Kim, D.H., Heterojunction C3N4/MoO3 microcomposite for highly efficient photocatalytic oxidation of Rhodamine B. *Appl. Surf. Sci.*, 511, 2, 4, 5, 2020, doi: 10.1016/j.apsusc.2020.145595.
82. Hasija, V., Raizada, P., Sudhaik, A., Singh, P., Thakur, V.K., Khan, A.A.P., Fabrication of Ag/AgI/WO3 heterojunction anchored P and S co-doped graphitic carbon nitride as a dual Z scheme photocatalyst for efficient dye degradation. *Solid State Sci.*, 100, 2, 6, 9, 2020, doi: 10.1016/j.solidstatesciences.2019.106095.
83. Das, R.S., Warkhade, S.K., Kumar, A., Gaikwad, G.S., Wankhade, A.V., Graphitic carbon nitride @ silver zirconate nanocomposite (gC3N4@Ag2ZrO3): A Type-II heterojunction for an effective visible light photocatalysis and bacterial photo-inactivation. *J. Alloys Compd.*, 846, 2, 7, 9, 2020, doi: 10.1016/j.jallcom.2020.155770.
84. Sivasakthi, S. and Gurunathan, K., Graphitic carbon nitride bedecked with CuO/ZnO hetero-interface microflower towards high photocatalytic performance. *Renewable Energy*, 159, 791, 793, 795, 796, 799, 819, 826, 828, 2020, doi: 10.1016/j.renene.2020.06.027.
85. Kuila, S.K., Sarkar, R., Kumbhakar, P., Kumbhakar, P., Tiwary, C.S., Kundu, T.K., Photocatalytic dye degradation under sunlight irradiation using cerium ion adsorbed two-dimensional graphitic carbon nitride. *J. Environ. Chem. Eng.*, 8, 4, 1, 2, 6, 7, 9, 2020, doi: 10.1016/j.jece.2020.103942.
86. Shaheen, N. *et al.*, Carbon coated tungsten doped molybdenum oxide nanowires and their composite with graphitic carbon nitride for photocatalysis and antibacterial studies. *Ceram. Int.*, 49, 4, 6906–6922, 2023, doi: https://doi.org/10.1016/j.ceramint.2022.10.076.

2

Nanocomposites for Pharmaceutical Waste Degradation: A Remediation Tool

Sharmila Suresh[1], Pragna M. Shivannanavar[1], Ravikumara Moodalakoppalu Yogarajachari[1], Shankramma Kalikeri[1*], Charan Kumar Kachintaya[2] and Lingaraju Honnur Gurusiddappa[3]

[1]Division of Nanoscience and Technology, School of Lifesciences, JSS Academy of Higher Education and Research Sri Shivarathreeshwara Nagara, Mysore, Karnataka, India

[2]#9A, Shankar Colony I Cross, SN Pet, Ballari, India

[3]Department of Environmental Science (School of Life Sciences), JSS Academy of Higher Education and Research, Mysore, India

Abstract

Contaminants of emergent concern such as pharmaceutical composites have sparked a widespread concern due to their persistence, bioaccumulation, and toxicity. Human and animal consumption has been identified as the primary sources of medication contamination. Traditional methods of treating medicinal substances are ineffective at degrading them. Advanced wastewater treatment technologies are being researched to properly remove pollutants, particularly pharmaceutical chemicals. Photocatalysis is one of many applications, where nanocomposites have proven to be strong contenders. The current chapter covers a wide range of nanocomposites that could be used in environmental treatments; photocatalysis reduced the toxicity of pharmaceutically active compounds while improving wastewater effluent biodegradability.

Keywords: Pharmaceutical waste, advanced oxidation processes (AOPs), photocatalysis, nanocomposites, degradation

**Corresponding author*: shankrutk@jssuni.edu.in; shankrutk@gmail.com

Sadanand Pandey, Elvis Fosso-Kankeu and Soumya Pandit (eds.) Synergy of Bio-Chemical Processes for Photocatalytic and Photoelectrochemical Wastewater Treatment, (27–48) © 2024 Scrivener Publishing LLC

2.1 Introduction

Wastewater from pharmaceutical companies and homes contains a variety of drugs, including antibiotics and other components, and it enters the aquatic system freely. As a result, the life-threatening effects of pharmaceutical medicines on aquatic life and all living species in each ecosystem are being passed through the municipal sewage system without being recycled (Kumar *et al.*, 2022). Moreover, medicines are often dumped directly into the environment, and pharmaceutical wastes from the manufacturing industry are disposed straight into water sources (Lin *et al.*, 2014). Non-steroidal anti-inflammatory drugs (NSAIDs) are highly water-soluble medications used to treat a wide range of disease condition, including inflammatory diseases and pain relief. The process of disintegration is slow in NSAIDs, which, in, turn results in the harmful effect on living species and environment (Hirsch *et al.*, 1998). Historically, a number of conventional techniques were used for the degradation of pharmaceutical waste. Primarily, they are categorized into three ways: chemical physical, and biological (Hirsch *et al.*, 1998). In general, physical methods, such as membrane filtration and adsorption, are very effective means of removing pharmaceutical waste from wastewater. Adsorption is a straightforward and efficient technique for the elimination of pollutants (Kam *et al.*, 2018). Pressure moving across a membrane also drives the filtration process, and the benefits of using filtration include a higher rejection rate than ultrafiltration and a higher flux than reverse osmosis. However, filtration membrane fouling is a disadvantage of this method. Chemical reagents or radicals created chemically are used as major components in chemical operations to oxidize or decompose waste pollutants. Such procedures include "coagulation, flocculation, ozonation, ion exchange, and advanced oxidation." Photocatalytic degradation is more efficient method for degradation because, when a semiconductor is utilized as a photocatalyst, its surface is first irradiated with a certain wavelength of light. "When exposed to light with an energy equal to or greater than the bandgap energy, electrons from the valance band are stimulated into the conduction band, while holes stay in the valance band (Kalikeri *et al.*, 2018b; Rajasri *et al.*, 2022)." "The process of introducing coagulants into a medium in order to destabilize the particles is known as coagulation" (Abinaya *et al.*, 2019). Increasing their proclivity to form agglomerates, followed by flocculation, allows for the creation of bigger agglomerates. But toxicity is a major problem in this method. Microorganisms are used in biological processes to break down and adsorb pharmaceutical contaminants in wastewater.

Plants are commonly used in biological approaches such as phytoremediation and microorganism bioremediation. The techniques employed by microorganisms to remove colors are aerobic, anaerobic, or a combination of both. Biological treatments are still utilized because they are affordable and can handle a wide range of organic wastes. In biological method, it takes time for the organisms to decay, and some chemical substances, such as dyes, are resistant to aerobic treatment. "To remove organic contaminants, modern wastewater treatment methods such as membrane-based processes, advanced oxidation processes (AOPs), and UV-lighted treatments are used" (Kalikeri and Kodialbail, 2023a; Rajasri *et al.*, 2022). The key benefit of AOPs is that they successfully degrade aquatic pollutants using highly reactive HO (hydroxyl radicals), whereas basic approaches and activated carbon-based adsorption techniques mostly entail physical pollution modification with no degradation (Gopalakrishnan *et al.*, 2023; Villeneuve *et al.*, 2014). Several AOPs, including "UV/H_2O_2, ozone-based, Fenton-based, and semiconductor-based photocatalytic processes," are used to effectively degrade organic contaminants. Due to the lower input of chemical components and the utilization of oxygen from the air as an oxidant, semiconductor-based photocatalytic processes are commonly applied in comparison to other AOPs that need relatively expensive oxidizing agents such as "H_2O_2, O_3, TiO_2, CeO_2, ZnO, CdS, WO_3, and Fe_2O_3," and a variety of other semiconductor nanostructures are used as photocatalysts.

2.2 Pharmaceutical Waste Degradation

The importance of safe drinking water and the finding of sufficient water sources for daily usage cannot be emphasized; the majority of our clean water resources are used for daily household purposes from dawn to dusk, as well as other institutions such as hospitals, educational institutions, and working places and offices, while the remaining semi-pure water resources are used for crop cultivation in farms, industrial sites for production, and animal rearing. The pharmaceutical waste enters the water supply; it begins to impair aquatic life and the health of water-consuming organisms. The use of birth control pills and their metabolites is a major source of worry because they pollute the water supply and have a significant influence on watery things such as living organelles, fish, and animals that are part of the water environment (Chen, 2020). The ineffective removal of those pharmaceutical drugs is required, which may cause an adverse impact on the human body, where both men and women have a possibility of infertility, and the physical influence can be passed on to future generations. We study the most common pharmaceutically active compounds such as antibiotics, NSAIDs, steroids, and hormones, out of the 7,000 pharmacological prescriptions administered, due to their uncontrolled use and frequent presence in aqueous surfaces (Mohamed Isa *et al.*, 2021). Pharmaceutical wastewater is not only complex and concentrated, but it also contains several pollutants that are difficult to break down. Hence, a lot of conventional techniques have been applied as follows.

Coagulation sedimentation: This method largely minimizes the number of contaminants in wastewater by successfully degrading pollutants in water, thereby accomplishing the purpose. The fundamental goal of this treatment method is the generation of a large volume of contaminates, while its main benefit is the high removal rate of salt, ammonia, and nitrogen content (Amin *et al.*, 2021). The limitation of this method is only reduction of salts, and some of the chemicals do not undergo complete degradation.
Flotation: Flotation is often referred to as air flotation. Pollutants in similar quantities in the wastewater bind together and float to the water surface, and, lastly, the solid and liquid in the wastewater are efficiently separated, minimizing water pollution fundamentally (Kyzas and Matis, 2019). As such the fundamental restriction of this approach is that only separation of liquid and solid waste takes place, this method of degradation is partial.
Membrane separation: The membrane, which separates the solvent, is a critical component of the membrane separation process. "This technology effectively intercepts and recovers ethanol from polyphenol medicinal effluent" (Liang *et al.*, 2019). The drawback of this method is that it only recovers the ethanol from phenol compounds.
Electrolysis: "Medical sewage is subjected to a chemical reaction while being operated by an electric current. Compared to other methods, electrolysis has many advantages, including high efficiency, simplicity of use, and superior decolorization results" (Grigoriev *et al.*, 2020). The limitation of this method is that it requires continues supply electricity without which electrolysis will not be carried out.
Ultrasonic treatment: The three methods that used ultrasonic waves are supercritical water oxidation, radical oxidation, and combustion and breakdown in bubbles, and OH is at the core of ultrasonic water treatment. Researchers are becoming increasingly interested in the use of microwave and ultrasonic treatment tools in the field of pharmaceutical water treatment. Microwave treatment, particularly the combined process of biological contact oxidation to purify high-concentration and ultrasonic organic wastewater, is now commonly used. The impact of treatment has created a

big change in the process (Xu *et al.*, 2019). The drawback of this procedure is oxidation and combustion that take place, which result in releasing the pollutants to the atmosphere.

Ozonation: "The concentration of dissolved oxygen (DO) in the effluent is increased by ozonation. Increased DO can reduce the need for reaeration while raising the DO level in the receiving stream. Some viruses, spores, and cysts may not be inactivated efficiently at low concentrations." The disadvantage of this technology is that it is expensive and difficult to store and transport, and ozone is a hazardous gas that causes health problems.

Fenton Process: Fenton is one of the most efficient treatment methods for reducing a wide range of toxic organic contaminants from wastewater. Despite its many benefits, Fenton has a number of downsides, including a harsh acid environment, a high reagent cost, and a substantial amount of ferric sludge generation, which restricts its further use. One of the downsides of the Fenton process is the production of significant quantities of ferrous iron sludge, a limitation that might be circumvented by the use of nanoscale zero-valent iron (nZVI) powder. However, the methods that are mentioned do not entirely mineralize the contaminants, generate extra waste, and are energy-intensive. Therefore, nanotechnology, namely, AOPs/nano photocatalysis, is now gaining prominence. In contrast to traditional techniques that transfer contaminants from one medium to another, this procedure results in the production of harmless substances (Kalikeri *et al.*, 2018).

Nanotechnology has the potential to infiltrate through adsorption or photocatalytic degradation. The essential requirements for utilizing nanotechnology and nanoparticles are economically beneficial, user-friendly, and environmentally safe (Durairaj *et al.*, 2014). Photocatalysis can effectively and completely convert organic contaminants into inorganic substances; therefore, its application potential for decontaminating liquid waste is greater (Rajasri *et al.*, 2022). When light energy penetrates photocatalytic materials and sufficient energy falls on the surface, electrons and holes may be excited and created. The produced electron will react with the surrounding oxygen to produce superoxide anion (Rajasri *et al.*, 2022). After releasing electrons, photocatalytic materials become positively charged and react with moisture transformed into a hydroxyl radical ($OH-$). The produced superoxide and hydroxide radicals may decompose or destroy organic materials and microbiological components such as viruses, bacteria, fungus, and odors, among other things. Thus, photocatalytic materials contribute to environmental purification (Kalikeri *et al.*, 2018).

AOPs are distinguished by the production of large quantities of hydroxyl radicals that oxidize the bulk of complex molecules in effluent water (Kalikeri *et al.*, 2018). "The AOP technique is extremely effective at removing biologically toxic or non-degradable pollutants from wastewater, such as aromatics, pesticides, petroleum components, and volatile organic compounds" (Kalikeri *et al.*, 2018a). AOPs can also be used to treat the effluent from secondary treated wastewater, a process known as tertiary treatment. Mineralization occurs when contaminating materials are mainly transformed into stable inorganic substances such as water, carbon dioxide, and salts. "The goal of AOP wastewater purification is to minimize toxicity to the point that the cleansed wastewater can be returned to receiving streams or, at the very least, into a standard sewage treatment facility" (Valdivia *et al.*, 2023).

So far, the most prevalent method is photocatalysis, which has its own importance owing to the reusability of the photocatalyst, making it cost-effective and environmentally friendly. Nano-based photocatalysts are among those that are simple to produce and may efficiently degrade pollutants (Kalikeri *et al.*, 2023). The ZnO, TiO_2, and CeO_2 NPs are the most prevalent and effective nano photocatalysts used in this application. "A photocatalyst's efficiency is determined by its surface-active sites, band edge location, narrow bandgap energy, low rate of charge recombination, accelerated charge separation, and charge transfer. Limitation is large amount of residue in environment and light source play a major role" (Khan *et al.*, 2019). Table 2.1 lists the nanomaterials that are used in the degradation of pharmaceutical waste.

Amoxicillin is the common antibiotic that is prescribed by the doctor, which reaches the aquatic bodies by urine and excreta, which, in turn, starts affecting aquatic life further toxicity spreads to the other organisms. According to Li *et al.* (2018) TiO_2 nanoparticles were prepared by solvothermal method, and the particle shape is irregular spherical with the particle size of 10 nm. The degradation was done with the pharmaceutical drug Amoxicillin and TiO_2 nanoparticles in the presence of UV light; the amount of degradation shown was 96.7%. Erythromycin is the pharmaceutical antibiotic drug, which causes the toxicity that imbalances the ecosystem and also creates the disabling consequences in aquatic organisms. Roy and Chakraborty (2021) in their work stated that ZnO was prepared by hydrothermal method, which showed the particle structure to be nanorods with the particle size of 13.55 nm. ZnO was used to degrade the erythromycin drug in the presence of sunlight, and the percentage of degradation obtain was 88.8%. Streptomycin drug is an antibiotic that causes the toxicity to the fresh water algae; it also blocks the electron transportation and

Table 2.1 Summary of the nanoparticles used to degrade pharmaceutical waste.

Nanoparticles; bandgap (eV) Concentration ($mg\ L^{-1}$)	**Method of preparation**	**Shape; size**	**Drug/concentration ($mg\ L^{-1}$)**	**% Dye degradation; time (min)**	**Light source**	**References**
WO_3; 3.21-eV Conc; 2 mg	Precipitation method	Crystal; 8.1 nm	Lidocaine (250 mg)	97.5% (60 min)	Sunlight	(Fakhri and Behrouz, 2015)
TiO_2; 3.4-eV Conc; 2 mg	Solvothermal method	Irregular spherical; 10 nm	Amoxicillin (5 mg)	96.7% (60 min)	UV light	(Li *et al.*, 2018)
AgNPs; 2.33-eV Conc; 5 mg	Biological method	Spherical; 0–50 nm	Norfloxacin (5 mg)	80% (120 min)	Sun light	(Kanagamani *et al.*, 2019)
$TiO_{2;}$ 1.76-eV Conc; 10 mg	Sol-gel method	Crystal; 5 nm	Ciprofloxacin (5 mg)	78% (60 min)	Sun light	(Parmar and Srivastava, 2021)
ZnO; 3.2-eV Conc; 5 mg	Hydrothermal method	Nanorods; 13.55 nm	Erythromycin (5 mg)	88.8% (120 min)	Solar light	(Roy and Chakraborty, 2021)
CeO_2; 3.1-eV Conc; 15 mg	Nitrate combustion method	Regular spherical shapes; 2 nm	Streptomycin (15mg)	72% (90 min)	Visible light	(Iqbal *et al.*, 2022)

also ROS overproduction. It mainly affects the photosynthetic organisms in the aquatic environments by inhibiting protein synthesis. Iqbal *et al.* (2022) have synthesized CeO_2 nitrate by the combustion method, and the obtained particle size was 2 nm with the particle structure of spherical. The process of degradation was done with the drug streptomycin and CeO_2 nitrate nanoparticle in the visible light, and the degradation obtain was 72%. Lidocaine is a local anesthetic drug that does not undergo the biodegradation, and the metabolism of aquatic organisms is changed. Fakhri and Behrouz (2015) have carried an experiment by synthesized WO_3 by precipitation method, with a particle size of 8.1 nm with the particle structure of crystallite. When the degradation was carried out between the drug and WO_3 nanoparticle in the presence sunlight, the percentage degradation was shown to be 97.5%. Norfloxacin drug is an antibiotic; it then causes toxicity by inhibiting the dissipation, lengthening the half-lives, and enhancing residues of herbicides in sediment that increases the risk in aquatic bodies. Kanagamani *et al.* (2019) in their work synthesized silver nanoparticle by biology method with a particle size around 50 nm, and the particle structure was spherical. The process of degradation was carried out between norfloxacin drug and silver nanoparticle in the presence of sunlight, and the amount of degradation obtained was 80%. Ciprofloxacin drug is an antibiotic, causes the toxicity by making the bacterial resistant to the drug and also change the composition of single celled communities. Parmar and Srivastava (2021) in their work stated that TiO_2 was prepared by sol-gel method, with the crystalline structure. TiO_2 was used to degrade ciprofloxacin drug in the presence of sunlight, and the resultant percentage degradation was 78%.

Nanoparticles are employed to remove colors from industrial effluent. However, nanoparticles have certain limits in dye degradation. The agglomeration of nanoparticles is one of the key restrictions because it limits the surface area of the nanoparticles, which, in turn, reduces their efficacy in dye degradation. Another drawback is that, if the nanoparticles are not removed from the water, then they might be hazardous to aquatic life (Kalikeri *et al.*, 2018a; Kalikeri and Kodialbail, 2023b). Nanocomposites are better for degradation when compared to nanoparticles because they have a large surface area and can be more easily degraded by microbes (Kalikeri *et al.*, 2018b; Kalikeri and Kodialbail, 2023c). "Nanocomposites are more reactive because they have a high surface area compared to nanomaterial" (Remya *et al.*, 2022), and these have efficient rate degradation of pharmaceutical waste. Table 2.2 shows the list of nanocomposites used for pharmaceutical waste degradation.

Table 2.2 List of nanocomposites used for pharmaceutical waste degradation.

Nanocomposite bandgap (eV); Concentration (mg L^{-1})	Method; size; shape	Drug/ concentration (mg L^{-1})	% Dye degradation/ time (min)	Light sources	Reference
TiO_2/rGO/g-C_3N_4	Precipitation; crystal; 10nm	Tetracycline	79.5%	Visible light	(Abinaya *et al.*, n.d.2019)
Nickel ferrite/ chitosan/ bismuth (III) oxoiodide	Chemical; precipitation; cluster; 40 nm	Metronidazole	69.3% (200 min)	Sunlight	(Bhuyan *et al.*, n.d.2019)
Cadmium-CdS/ $Bi_4V_2O_{11}$; 2.4-eV Conc; 0.02 g	Chemical precipitation; crystal; 10–40nm	Ciprofloxacin; 10 mg	76.97% (40 min)	Visible light	(Attarikhasraghi *et al.*, 2021a)
Calcium-$CaCo_3$; 1.93-eV Conc; 2 mg	Electroporation; nanowire; 60 nm	Tetracycline hydrochloride; 15 mg	82% (40 min) 40% (120 min)	Sunlight Visible light	(Ban *et al.*, 2021)
Cobalt-$CoFe_2O_4$; 1.46-eV Conc; 2 μM	Thermal decomposition; cluster; 106 nm	Ciprofloxacin; 10 mg	94.38% (60 min)	Sunlight	(Mmelesi *et al.*, 2021)
Strontium-$SrTiO_3$; 3.25-eV Conc; 5 μM	Chemical precipitation Spherical 30 nm	Sulfamethoxazole; 5 mg	48% (60 min)	Visible light (xenon lamp)	(Zhang and Liu, 2021)
		Enrofloxacin hydrochloride; 10 mg	91.1% (90 min)		
		Oxytetracycline danofloxacin; 10 mg	91.5% (120 min)		
		Levofloxacin; 5 mg	88.6% (90 min)		
Graphitic-C_3N_4; 2.6-eV Conc; 1.3 μM	Electroporation; nanosheets; 20 nm	Ciprofloxacin tetracycline; 5 mg	89% (2,400 min) 96% (240 min)	Sunlight	(Nasraoui *et al.*, 2021)
AgI/$Bi_2Sn_2O_7$	Ball milling; crystal; 10 nm	Tetracycline	83% (50 min)	Visible light	(Gulati *et al.*, 2022)

(*Continued*)

Table 2.2 List of nanocomposites used for pharmaceutical waste degradation. (*Continued*)

Nanocomposite bandgap (eV); Concentration (mg L^{-1})	Method; size; shape	Drug/ concentration (mg L^{-1})	% Dye degradation/ time (min)	Light sources	Reference
WS_2/PANI; 1.0-eV Conc; 500 μM	Intercalation-exfoliation method; nanosheets; 21.6 nm	Nitrofurantoin; 5 mg	95.38% (120 min)	Sunlight	(Fatima *et al.*, 2022)
BIOCHAR; 4.4-eV Conc; 0.01 g	Ball milling; crystal; 10 nm	Magnesium chloride hexahydrate; 5 mg	80.26% (90 min)	Sunlight	(Venkatesh *et al.*, 2022)
Bismuth-$Bi_{12}O_{17}Cl_{12}$; 2.1-eV Conc; 3 mg	Sol-gel method; nano powder; 20 nm	Sulfamethazine; 5 mg	90.9% (120 min)	Visible light	(Gulati *et al.*, 2022)
$NiFe_{12}O_{19}$-ZnO;3.37-eV Conc; 1 μM	Hydrothermal process; nanorods; 3 μm	Ceftriaxone sodium; 5 mg	89.34% (120 min)	Sunlight	(Javadmoosavi *et al.*, 2023)

Javadmoosavi *et al.* (2023) reported that $NiFe_{12}O_{19}$-ZnO nanocomposite for the degradation of the drug ceftriaxone (CFX). CFX is a third-generation cephalosporin antibiotic; this antibiotic was used to treat gonorrhoea, leptospirosis, and gram-negative bacteria. Photocatalytic processes are critical due to their high efficiency and short reaction time. When CFX was degraded using $NiFe_{12}O_{19}$-ZnO nanocomposite in the presence of sunlight that showed the 84% of percentage degradation in the time duration of 120 min. Generally, photocatalysis is influenced by photons and eventually results in the formation of free radicals. Antibiotics can be removed from the environment by these free radicals.

Venkatesh *et al.* (2022) stated that biochar is a nanosized substance with improved physical, chemical, and surface qualities. Biochar's high specific surface areas have depended on the parameter such as pH, temperature, mixer ratio, and H_2O_2. Venkatesh *et al.* (2022a) carried out a photocatalytic study of new composites of biochar to remove tetracycline (T.C.) exposed to varied circumstances. "At 25°C, the photodegradation efficiency was determined to be 80.26% for TC concentration in the presences of sunlight with 50 ppm, H_2O_2 of 100 mM, and pH 5–6 of Ag/MgO/biochar (0.01 g)."

This study enables the treatment of diverse organic wastewaters using metal oxide/biochar nanocomposites with significant adsorption and photocatalytic degradation capabilities.

Fatima *et al.* (2022) reported WS_2/PANI nanocomposite for nitrofurantoin (NFT) degradation; here, they fabricated WS_2 nanosheets using hydrothermal method, and an *in situ* polymerization technique and ammonium persulfate as an oxidizer were used to create the WS_2-PANI nanocomposite. NFT is a well-known antimicrobial drug related to NFT that belongs to the nitrofuran family and can impede the growth of a wide range of gram-positive and gram-negative bacteria. This antibiotic is widely used in veterinary medicine to treat urinary tract infections. Despite its promising antibacterial efficacy, an overdose can result in significant adverse effects including lung toxicity, nausea, neuropathy, and headache. This drug has been banned in several countries due to its carcinogenic and mutagenic properties. Thus, photodegradation studies demonstrated under sunlight that the photocatalytic activity of the WS_2/PANI composite over WS_2 for NFT is 95.4% in 120 min and WS_2 is 79% in 120 min. Because of its larger pore size and higher charge separation efficiency than pure WS2, WS2-PANI performed better as a photocatalyst on NFT (Fatima *et al.*, 2022). Figure 2.1 shows the photocatalytic mechanism in the degradation of pharmaceutical waste.

According to Gulati *et al.* (2022), sulfadimidine degradation was demonstrated, and this is one of the most widely used drugs in veterinary

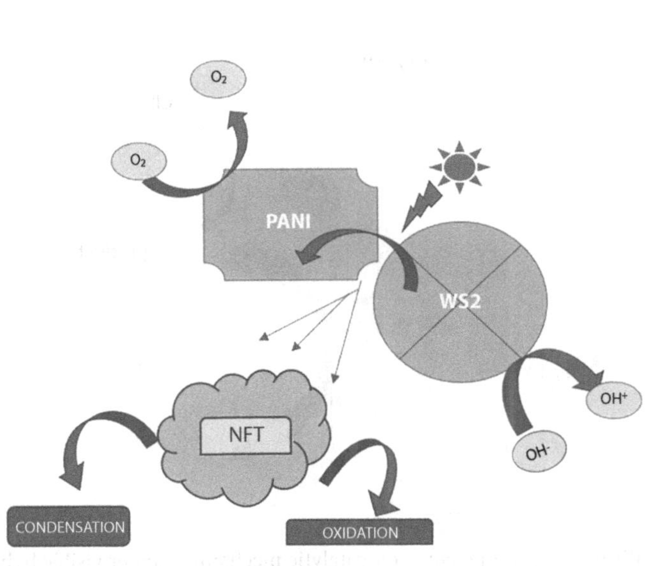

Figure 2.1 Photocatalytic degradation diagram (Fatima *et al.*, 2022).

medicine, particularly in pig production, where it is used to combat respiratory disease. The amount of AgI in the AgI/$Bi_{12}O_{17}Cl_{12}$ composites was found to play an important role in charge carrier separation, redox ability, and photostability of the system, with a 25% optimized ratio obtained, in the presence of sunlight. According to Gulati *et al.* (2022), the AgI/$Bi_2Sn_2O_7$ Z-scheme nano junction hybrid demonstrated superior photocatalytic performance for tetracycline removal, degrading approximately 83.0% of tetracycline within 50 min (500-W Xe lamp). Figure 2.2 shows the photocatalytic mechanism of Bi/Bi_3NbO_7 composite under visible light.

According to Attarikhasraghi *et al.* (2021), ciprofloxacin is an antibiotic used to treat bacterial infections. It is also used to treat urinary tract, nose, throat, skin, and soft tissue infections, as well as pneumonia (infection of the lungs). It stops bacterial cells from replicating and mending themselves, effectively destroying them. Developing cadmium alloy Z-scheme–based heterojunction photocatalytic devices to breakdown antibiotics is a viable method. The direct Z-scheme CdS/$Bi_4V_2O_{11}$ photocatalyst developed was found to have greater redox ability than pure CdS and $Bi_4V_2O_{11}$ samples. Under visible light, the CdS/$Bi_4V_2O_{11}$ (3:1) composite shows the grater degradation efficacy for ciprofloxacin (76.97%, 120 min).

Wu *et al.* (2015) mentioned that tetracycline is an antibiotic medication used to treat bacterial infections in the body. It is useful in treating infections of the lungs, urinary tract, eyes, and other organs (Ban *et al.*, 2021). $CaFe_2O_4$

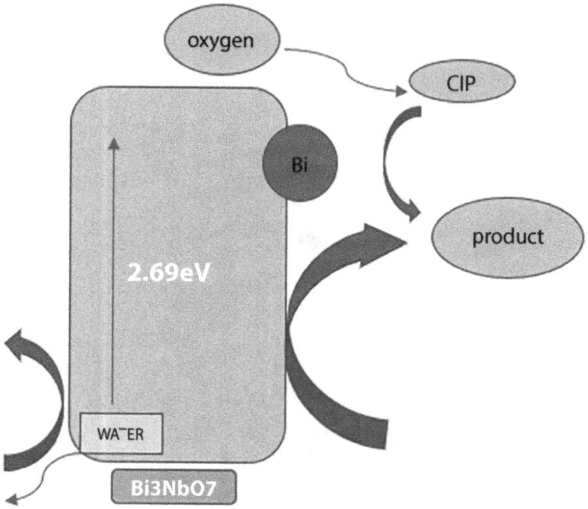

Figure 2.2 Bi/Bi_3NbO_7 composite photocatalytic mechanism under visible light (Gulati *et al.*, 2022).

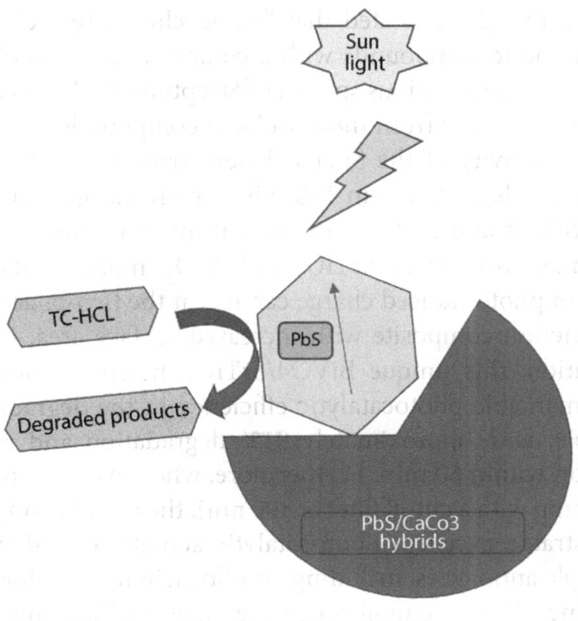

Figure 2.3 Photocatalytic degradation mechanism (Ban *et al.*, 2021).

nanowire-based nanocomposite exhibited higher photocatalytic degradation of tetracycline (40%) in visible light. The heterostructure of $CaFe_2O_4$ nanowires efficiently aligned molecular energy band locations, limiting photo-generated electron and hole recombination. Carbonate radicals also played important roles in the degradation of antibiotics in this regard.

Ban *et al.* (2021) and Javid *et al.* (2020) are using an egg/shell-based Pubs/$CaCO_3$ composite; carbonate radical (CO_3)–mediated photocatalytic tetracycline hydrochloride degradation was conducted under sunlight. This composite shows remarkable photocatalytic degradation efficiency of 82% for tetracycline after 40 min. Figure 2.3 shows the degradation of tetracycline hydrochloride using Pubs/$CaCO_3$ composite.

Mmelesi *et al.* (2021) derived that ciprofloxacin can have major adverse effects such as tendon difficulties, nerve damage, severe mood or behavior changes, and low blood sugar. The surface imprinting method was employed to create a stable core-shell imprinted silver-(poly-o-phenylenediamine)/CoFe2O4 spherical core-shell with excellent magnetic separation for ciprofloxacin degradation. The addition of silver-(poly-o-phenylenediamine) to the imprinted layer boosted photocatalytic activity dramatically, resulting in 94.38% ciprofloxacin degradation in 90 min under 250-W xenon lamp.

Zhang and Liu (2021) stated that "tetracyclines are a class of broad-spectrum antibiotic compounds with a common basic structure that are obtained directly from various species of Streptomyces bacteria or synthesized semi-synthetically from those isolated compounds." The observable photocatalytic activity of the $SrTiO_3$ heterostructure photocatalysts was 85% tetracycline degradation in 140 min, which was greater than that of the pristine $SrTiO_3$ and $Bi2O_3$. The efficient interface contact and heterojunction formation between $SrTiO_3$ and Bi_2O_3 improved the separation and transfer of photoinduced charge carriers at the two-phase interface of the heterojunction composite with increased surface area. Under xenon lamp irradiation, this unique BiVO4/SrTiO3 heterojunction composite displayed remarkable photocatalytic efficiency in the degradation of sulfamethoxazole, with approximately 91% degradation and roughly 48% mineralization within 60 min. Furthermore, when exposed to visible light (250-W Xe lamp with a cutoff filter at 400 nm), the CDs/$SrTiO_3$ heterojunction demonstrated excellent photocatalytic activity toward the degradation of multiple antibiotics, including enrofloxacin hydrochloride (91.1%), oxytetracycline (90.3%), danofloxacin mesylate (91.5%), and levofloxacin (88.6%).

According to Nasraoui *et al.* (2021), the antibiotics that are commonly prescribed are tetracycline and ciprofloxacin that have hazards issues when lead to the water body, which can be degraded by the ultrathin MoS_2/graphitic-C_3N_4 (MoS_2/g-C_3N_4) hybrid composite that was manufactured by creating interface contacts between ultrathin MoS2 nanosheets and g-C_3N_4. The hybrid MoS2/g-C3N4 photocatalyst containing 5% MoS2 demonstrated superior ciprofloxacin and tetracycline removal, with 81.9% and 96.0% ciprofloxacin and tetracycline degradation, respectively, in 240 min under sunlight.

Abinaya *et al.* (2019) in their work mentioned that TiO_2/rGO/g-C_3N_4 nanocomposites were synthesized by precipitation method; the particle size was 10 nm, and the shape is crystal. The degradation was carried on the drug tetracycline under visible light; the percentage of degradation obtained was 79.5%. Bhuyan *et al.* (2019) in their work reported that nickel ferrite/chitosan/bismuth (III) oxoiodide nanocomposites were synthesized by chemical precipitation; the particle size was 40 nm and the shape is cluster. The degradation was carried on the drug metronidazole in the presence of sunlight; the obtained percentage degradation was 60.3% in the time duration of 200 min. Send *et al.* (2021b) reported that cadmium-CdS/$Bi_4V_2O_{11}$ nanocomposites were prepared by chemical precipitation method; the obtained particle was 10–40 nm in size and crystal in structure. The degradation was carried on the ciprofloxacin in the presence of visible time, and the obtained percentage degradation was 76.97% in the time interval of 40 min.

Ban *et al.* (2021) in their work stated that calcium ($CaCo_3$) nanocomposites were synthesized by electroporation; the obtained particle size was 60 nm, and the structure was nanowires. Tetracycline drug was degraded using calcium nanocomposite, and the obtained percentage degradation was 82% in sunlight and 40% in visible light for the duration of 20 min and 40 min, respectively. Mmelesi *et al.* (2021) in their work reported that $CoFe_2O$ composites were synthesized by thermal decomposition method; the particle size was 106 nm, whereas the structure was in cluster. The degradation was taken place between the ciprofloxacin drug and cobalt nanocomposites in the presence of sunlight; the percentage degradation achieved was 94.38% in the time interval of 60 min; they stated that $SrTiO_3$ nanocomposites were prepared by chemical precipitation method that showed the particle size of 30 nm and the shape is spherical. The degradation was taken place between four drugs: sulfamethoxazole, enrofloxacin hydrochloride, oxytetracycline, and levofloxacin; the degradation was carried in the presence of xenon lamp; the percentage degradation achieved was 48%, 91.1%, 91.5%, and 88.6% in the time interval of 60 min, 90 min, 120 min, and 90 min, respectively.

Kumar *et al.* (2022) reported in their work that graphitic-C_3N_4 was synthesized by electroporation method; the particle achieved was 20 nm with a nanosheet shape. The degradation was taken place with drugs ciprofloxacin and tetracycline; the obtained percentage degradation was 89% and 96%, respectively, under sunlight with the time duration 240 min. Gulati *et al.* (2022) in their work explained that $AgI/Bi_2Sn_2O_7$ composite was synthesized by ball milling method; the obtained particle size was 10 nm with a crystal shape. The degradation was carried with drug tetracycline; the obtained percentage degradation was 83% in 50 min in the presence of visible light. Fatima *et al.* (2022) in their work mentioned that WS_2/PANI was synthesized by intercalation exfoliation method with the particle size of 21.6 nm with a nanosheet structure. The degradation was carried out with the drug NFT in the presence of sunlight; the obtained percentage degradation was 95.38% in the time interval of 120 min.

Venkatesh *et al.* (2022) in their work showed that biochar nanocomposite was synthesized by ball milling process; the obtained particle size was 10 nm with a crystal structure. The degradation was carried out on the drug magnesium chloride hexahydrate in the presence of sunlight; the obtained percentage degradation was 80.26% in the time duration of 90 min. Gulati *et al.* (2022) in their work reported that $Bi_{12}O_{17}Cl_{12}$ nanocomposites were synthesized by sol-gel method; particle size was 20 nm, with nano powder structure. The degradation was carried out with the drug sulfamethazine in the presence of visible light; the obtained percentage degradation was 90.9% in the time interval of 120 min. Javadmoosavi *et al.* (2023) in their work mentioned that $NiFe_{12}O_{19}$-ZnO was synthesized by hydrothermal method; the particle size was 3 nm with a nanorod structure. The degradation was carried out with the drug ceftriaxone in the presence of sunlight; the percentage degradation achieved was 89.34% in the time interval of 120 min.

Through these reports, the nanocomposites that showed more efficient results are cobalt, strontium, graphitic, and WS_2/PANI. These four nanocomposites show more than 90% of percentage degradation when compared to other composites.

2.3 Conclusion

With the increase in poor health conditions, doctors are prescribing more alternative drugs; the same drugs are available in different brands after the expiry time; a greater number of drugs are dumped and also thrown in water bodies without prier treatment, and it is difficult to mineralize pharmaceutical waste using traditional methods. So, AOP is the best option, in that nanocomposites for waste degradation are a very easy and effective method, and it is an eco-friendly process that produces good results. Thus, we can reclaim water and other resources with certainty by using nanocomposites. By increasing the use of nanocomposites, the majority of pharmaceutical waste can be degraded in the future; instead of dumping in soil or throwing in water bodies, we can degrade them using the nano-based materials that are cost-efficient, require less labor, and are less in time; hence, more degradation can be achieved.

2.4 Future Perspectives

The degradation of pharmaceutical waste will result in environmental cleansing by reducing contamination of water bodies. Aquatic fauna consumes pharmaceutical waste, which, in turn, causes human poisoning. Poisoning affects the food chain, and environmental disruption begins. Thus, nanoparticles used for degradation are the best option because they are eco-friendly and complete mineralization occurs, so they have no effect on the environment. Because of their unique properties, nanocomposite can be used for further research, and we can also change the light source in an efficient manner. The following are the primary benefits of photostatic pharma waste degradation:

- Nanocomposites degrade faster and in a more environmentally friendly manner.
- A large amount of waste can be degraded under sunlight.
- Nanocomposite degrades more quickly than nanoparticles.

References

Abinaya, M. and Rajakumaran, R., Performance of Polypyrrole Polymer-Incorporated Ag2MoO4 Nanocomposite for Detection and Degradation of Environmental Pollutants and Pharmaceutical Drugs, 2019, ACS Publications. *ACS Appl. Mater. Interfaces,* 11, 41, 38321–38335 Retrieved March 29, 2023, from https://pubs.acs.org/doi/abs/10.1021/acsami.9b13682, S. C.-A. applied, undefined. (n.d.).

Abinaya, M., Rajakumaran, R., Chen, S.M., Karthik, R., Muthuraj, V., *In Situ* Synthesis, Characterization, and Catalytic Performance of Polypyrrole Polymer-Incorporated Ag2MoO4 Nanocomposite for Detection and Degradation of Environmental Pollutants and Pharmaceutical Drugs. *ACS Appl. Mater. Interfaces, 11,* 41, 38321–38335, 2019, https://doi.org/10.1021/acsami.9b13682.

Amin, A., Al Bazedi, G., Abdel-Fatah, M.A., Experimental study and mathematical model of coagulation/sedimentation units for treatment of food processing wastewater. *Ain Shams Eng. J., 12,* 1, 195–203, 2021, https://doi.org/10.1016/J.ASEJ.2020.08.001.

Attarikhasraghi, N., Zare, K., Mehrizad, A., Modirshahla, N., Behnajady, M.A., Achieving the enhanced photocatalytic degradation of ceftriaxone sodium using CdS-g-C3N4 nanocomposite under visible light irradiation: RSM modeling and optimization Achieving the enhanced photocatalytic degradation of ceftriaxone sodium using CdS-g-C3N4 nanocomposite under visible light irradiation: RSM modeling and optimization. *J. Inorg. Organomet. Polym. Mater.*, 31,4, 3164–3174 2021a, https://doi.org/10.21203/rs.3.rs-212164/v1.

Ban, Q., Liu, T., Ning, K., Fan, J., Cui, Q., Guo, Y., Zai, X., Effect of calcium treatment on the browning of harvested eggplant fruits and its relation to the metabolisms of reactive oxygen species (ROS) and phenolics. *Food Sci. Nutr., 9,* 10, 5567–5574, 2021, https://doi.org/10.1002/FSN3.2517.

Bhuyan, A., Recent advances in new generation nanocomposite materials for adsorption of pharmaceuticals from aqueous environment. *Environ. Sci. and Poll. Res.,* 30, 39377–39417, 2023, Springer, Retrieved March 29, 2023, from https://link.springer.com/article/10.1007/s11356-023-25707-0, Pollution, M. A.-E. S. and, & undefined. (n.d.).

Chen, Y., Photodegradation of pharmaceutical waste by nano-materials as photocatalysts, in: *Nano-Materials as Photocatalysts for Degradation of Environmental Pollutants: Challenges and Possibilities*, pp. 143–152, 2020, https://doi.org/10.1016/B978-0-12-818598-8.00008-0.

Durairaj, S., Application of nanoparticles in wastewater treatment. *Pollut. Res.*, 33, 3, 567-571, Retrieved March 17, 2023, from https://www.researchgate.net/publication/316701373.

Fakhri, A. and Behrouz, S., Photocatalytic properties of tungsten trioxide (WO3) nanoparticles for degradation of Lidocaine under visible and sunlight irradiation. *Sol. Energy*, *112*, 163–168, 2015, https://doi.org/10.1016/J.SOLENER.2014.11.014.

Fatima, T., Husain, S., Khanuja, M., Superior photocatalytic and electrochemical activity of novel WS2/PANI nanocomposite for the degradation and detection of pollutants: Antibiotic, heavy metal ions, and dyes. *Chem. Eng. J. Adv.*, 12, 6, 7, 2022, https://doi.org/10.1016/j.ceja.2022.100373.

Gopalakrishnan, G., Jeyakumar, R.B., Somanathan, A., Challenges and Emerging Trends in Advanced Oxidation Technologies and Integration of Advanced Oxidation Processes with Biological Processes for Wastewater Treatment. *Sustainability*, *15*, 5, 4235, 2023, https://doi.org/10.3390/SU15054235, 2023, Vol. 15, Page 4235.

Grigoriev, S.A., Fateev, V.N., Bessarabov, D.G., Millet, P., Current status, research trends, and challenges in water electrolysis science and technology. *Int. J. Hydrogen Energy*, *45*, 49, 26036–26058, 2020, https://doi.org/10.1016/J.IJHYDENE.2020.03.109.

Gulati, S., Goyal, K., Arora, A., Kumar, S., Trivedi, M., Jain, S., Bismuth ferrite (BiFeO3) perovskite-based advanced nanomaterials with state-of-the-art photocatalytic performance in water clean-up. *Environ. Sci. Water Res. Technol.*, *8*, 8, 1590–1618, 2022, https://doi.org/10.1039/D2EW00027J.

Hirsch, R., Ternes, T.A., Haberer, K., Mehlich, A., Ballwanz, F., Kratz, K.-L., Determination of antibiotics in different water compartments via liquid chromatography-electrospray tandem mass spectrometry. *In J. Chromatogr. A*, 815, 3, 4, 1998.

Iqbal, J., Shah, N.S., Khan, Z.U.H., Rizwan, M., Murtaza, B., Jamil, F., Shah, A., Ullah, A., Nazzal, Y., Howari, F., Visible light driven doped CeO2 for the treatment of pharmaceuticals in wastewater: A review. *J. Water Process Eng.*, *49*, 103130, 2022, https://doi.org/10.1016/J.JWPE.2022.103130.

Javadmoosavi, S.Y., Naghizadeh, A., Mizwari, Z.M., Mortazavi-Derazkola, S., Biosynthesis of novel NiFe12O19-X (X = ZnO and TiO2) magnetic nanophotocatalyst toward the degradation pharmaceutical ceftriaxone sodium from aqueous solution under sunlight irradiation and antibacterial activity. *Ceram. Int.*, *49*, 1, 1351–1361, 2023, https://doi.org/10.1016/j.ceramint.2022.09.116.

Javid, N., Honarmandrad, Z., Malakootian, M., *Ciprofloxacin removal from aqueous solutions by ozonation with calcium peroxide*, 2020, 2019. https://doi.org/10.5004/dwt.2020.24855.

Kalikeri, S., Kamath, N., Gadgil, D.J., Shetty Kodialbail, V., Visible light-induced photocatalytic degradation of Reactive Blue-19 over highly efficient polyaniline-TiO2 nanocomposite: a comparative study with solar and UV photocatalysis. *Environ. Sci. Pollut. Res.*, *25*, 4, 3731–3744, 2018a, https://doi.org/10.1007/S11356-017-0663-1/METRICS.

Kalikeri, S. and Kodialbail, V.S., Visible light active Bismuth ferrite embedded TiO2 nanocomposite structures for dye mineralization by photocatalysis -A strategy to harness solar energy for remediation of water contaminated with mixture of dyes. *Surf. Interfaces*, *36*, 102492, 2023a, https://doi.org/10.1016/J.SURFIN.2022.102492.

Kalikeri, S. and Kodialbail, V.S., Visible light active Bismuth ferrite embedded TiO2 nanocomposite structures for dye mineralization by photocatalysis -A strategy to harness solar energy for remediation of water contaminated with mixture of dyes. *Surf. Interfaces*, *36*, 102492, 2023b, https://doi.org/10.1016/J.SURFIN.2022.102492.

Kalikeri, S. and Kodialbail, V.S., Visible light active Bismuth ferrite embedded TiO2 nanocomposite structures for dye mineralization by photocatalysis -A strategy to harness solar energy for remediation of water contaminated with mixture of dyes. *Surf. Interfaces*, *36*, 102492, 2023c, https://doi.org/10.1016/J.SURFIN.2022.102492.

Kam, C.S., Leung, T.L., Liu, F., Djurišić, A.B., Xie, M.H., Chan, W.K., Zhou, Y., Shih, K., Lead removal from water-dependence on the form of carbon and surface functionalization. *RSC Adv.*, *8*, 33, 18355–18362, 2018, https://doi.org/10.1039/c8ra02264j.

Kanagamani, K., Muthukrishnan, P., Shankar, K., Kathiresan, A., Barabadi, H., Saravanan, M., Antimicrobial, Cytotoxicity and Photocatalytic Degradation of Norfloxacin Using Kleinia grandiflora Mediated Silver Nanoparticles. *J. Cluster Sci.*, *30*, 6, 1415–1424, 2019, https://doi.org/10.1007/S10876-019-01583-Y/METRICS.

Khan, I., Saeed, K., Khan, I., Nanoparticles: Properties, applications and toxicities. *Arabian J. Chem.*, *12*, 7, 908–931, 2019, https://doi.org/10.1016/J.ARABJC.2017.05.011.

Kumar, A., Sharma, G., Naushad, M., ALOthman, Z.A., Dhiman, P., Environmental Pollution Remediation via Photocatalytic Degradation of Sulfamethoxazole from Waste Water Using Sustainable Ag2S/Bi2S3/g-C3N4 Nano-Hybrids. *Earth Syst. Environ.*, *6*, 1, 141–156, 2022, https://doi.org/10.1007/s41748-021-00223-8.

Li, D., Dai, S., Li, J., Zhang, C., Richard-Plouet, M., Goullet, A., Granier, A., Microstructure and Photocatalytic Properties of TiO2–Reduced Graphene Oxide Nanocomposites Prepared by Solvothermal Method. *J. Electron. Mater.*, *47*, 12, 7372–7379, 2018, https://doi.org/10.1007/s11664-018-6677-8.

Liang, B., He, X., Hou, J., Li, L., Tang, Z., Membrane Separation in Organic Liquid: Technologies, Achievements, and Opportunities. *Advanced Mater.*, *31*, 45, 1806090, 2019, https://doi.org/10.1002/ADMA.201806090.

Lin, S., Yip, N.Y., Elimelech, M., Direct contact membrane distillation with heat recovery: Thermodynamic insights from module scale modeling. *J. Membr. Sci.*, *453*, 498–515, 2014, https://doi.org/10.1016/j.memsci.2013.11.016.

Mmelesi, O.K., Masunga, N., Kuvarega, A., Nkambule, T.T., Mamba, B.B., Kefeni, K.K., Cobalt ferrite nanoparticles and nanocomposites: Photocatalytic, antimicrobial activity and toxicity in water treatment. *Mater. Sci. Semicond. Process.*, *123*, 105523, 2021, https://doi.org/10.1016/J.MSSP.2020.105523.

Mohamed Isa, E.D., Shameli, K., Ch'ng, H.J., Che Jusoh, N.W., Hazan, R., Photocatalytic degradation of selected pharmaceuticals using green fabricated zinc oxide nanoparticles. *Adv. Powder Technol.*, *32*, 7, 2398–2409, 2021, https://doi.org/10.1016/J.APT.2021.05.021.

Nasraoui, S., Al-Hamry, A., Teixeira, P.R., Ameur, S., Paterno, L.G., Ben Ali, M., Kanoun, O., Electrochemical sensor for nitrite detection in water samples using flexible laser-induced graphene electrodes functionalized by CNT decorated by Au nanoparticles. *J. Electroanal. Chem.*, *880*, 114893, 2021, https://doi.org/10.1016/J.JELECHEM.2020.114893.

Parmar, N. and Srivastava, J.K., Degradation of pharmaceutical antibiotic (ciprofloxacin) by photocatalysis process using sol-gel based titanium dioxide nanoparticles. *Int. J. Chem. Reactor Eng.*, *19*, 9, 929–938, 2021, https://doi.org/10.1515/IJCRE-2021-0076/MACHINEREADABLECITATION/RIS.

Rajasri, S., Kalikeri, S., Balachandran, S., Gowthami, K., Thirunarayanan, G., Swaminathan, M., Muthuvel, I., Enhanced photodegradation of 2,4-dinitrophenol by n–p type TiO2/BiOI nanocomposite. *J. Indian Chem. Soc.*, *99*, 3, 100337, 2022, https://doi.org/10.1016/J.JICS.2021.100337.

Remya, R.R., Julius, A., Suman, T.Y., Mohanavel, V., Karthick, A., Pazhanimuthu, C., Samrot, A.V., Muhibbullah, M., Role of Nanoparticles in Biodegradation and Their Importance in Environmental and Biomedical Applications. *J. Nanomater.*, *2022*, 3, 6, 2022, https://doi.org/10.1155/2022/6090846.

Roy, N. and Chakraborty, S., ZnO as photocatalyst: An approach to waste water treatment. *Mater. Today Proc.*, *46*, 6399–6403, 2021, https://doi.org/10.1016/J.MATPR.2020.06.264.

Valdivia, M.T., Taggart, M.A., Pap, S., Kean, A., Pfleger, S., Megson, I.L., Photocatalytic metallic nanomaterials immobilised onto porous structures: Future perspectives for at-source pharmaceutical removal from hospital wastewater and potential benefits over existing technologies. *J. Water Process Eng.*, *52*, 103553, 2023, https://doi.org/10.1016/J.JWPE.2023.103553.

Venkatesh, R., Sekaran, P.R., Udayakumar, K., Jagadeesh, D., Raju, K., Bayu, M.B., Adsorption and Photocatalytic Degradation Properties of Bimetallic Ag/MgO/Biochar Nanocomposites. *Adsorpt. Sci. Technol.*, *2022*, 5, 2022, https://doi.org/10.1155/2022/3631584.

Villeneuve, D.L., Crump, D., Garcia-Reyero, N., Hecker, M., Hutchinson, T.H., LaLone, C.A., Landesmann, B., Lettieri, T., Munn, S., Nepelska, M., Ottinger, M.A., Vergauwen, L., Whelan, M., Adverse Outcome Pathway (AOP)

Development I: Strategies and Principles. *Toxicol. Sci.*, *142*, 2, 312–320, 2014, https://doi.org/10.1093/TOXSCI/KFU199.

Wu, J., Jiang, Y., Zha, L., Ye, Z., Zhou, Z., Ye, J., Zhou, H., Tetracycline degradation by ozonation, and evaluation of biodegradability and toxicity of ozonation byproducts. *8*, 5, 580–586, 2015, https://doi.org/10.1680/JEES.2013.0061. AU: Please provide journal title.

Xu, X., Cao, D., Wang, Z., Liu, J., Gao, J., Sanchuan, M., Wang, Z., Study on ultrasonic treatment for municipal sludge. *Ultrason. Sonochem.*, *57*, 29–37, 2019, https://doi.org/10.1016/J.ULTSONCH.2019.05.008.

Zhang, X. and Liu, Y., Integrated forward osmosis-adsorption process for strontium-containing water treatment: Pre-concentration and solidification. *J. Hazard. Mater.*, *414*, 125518, 2021, https://doi.org/10.1016/J.JHAZMAT.2021.125518.

3

Nanostructure Metal Sulfides for Photocatalytic Water Remediation

Shuvendu Tripathy[1], Medha Kiran Patel[2], Ipsita Bose[2], Soumya Pandit[3] and Santimoy Khilari[2,4]*

[1]*Vikash Degree College, Bargarh, Odisha, India*
[2]*Department of Chemistry, Guru Ghasidas Vishwavidyalay, Bilaspur, Chhattisgarh, India*
[3]*Department of Life Sciences, School of Basic Sciences and Research, Sharda University, Noida, Uttar Pradesh, India*
[4]*Department of Chemistry, University of Allahabad, Prayagraj, Uttar Pradesh, India*

Abstract
Water is an essential commodity for civilization, and, without it, life cannot run. Many anthropogenic activities generate wastewater of different types. Thus, the treatment of wastewater to supply reusable water becomes necessary. In this context, development of efficient, cost-effective water remediation technology is an urgent need. Photocatalytic degradation of organic (dyes, synthetic pesticides, etc.) and inorganic (heavy metals, ions, etc.) pollutants is an emerging sustainable technology to treat wastewater emitting from different sources. Solar-assisted photocatalysis with suitable photocatalysts made the photocatalytic water remediation more attractive. The semiconductor photocatalysts play a key role in this remediation process. In past decades, numerous inorganic and organic semiconductors and their composites have been developed for these purposes. Among inorganic semiconductors, metal sulfides are considered to be the most potential candidate for their intriguing physiochemical properties. The efficiency of the photocatalytic waste treatment relies exclusively on the photon capture and their effective utilization. Tailoring of the optoelectronic and textural properties can boost the degradation capacity of the photocatalyst. In this aspect, development of nanostructure metal sulfide photocatalysts comes into the limelight. The photoreactivity and catalytic efficiency can be tuned by exploring the shape, size, and crystal structure of

**Corresponding author*: santimoy@allduniv.ac.in

Sadanand Pandey, Elvis Fosso-Kankeu and Soumya Pandit (eds.) Synergy of Bio-Chemical Processes for Photocatalytic and Photoelectrochemical Wastewater Treatment, (49–72) © 2024 Scrivener Publishing LLC

nanostructure metal sulfides. Development of heterostructure metal sulfides also found to be a noteworthy protocol. Moreover, the design of a suitable catalytic system for practical application is very essential. In the present chapter, the fundamental principle of the metal sulfide–based photocatalysis, the design of solar active metal sulfides, and the insights on the physiochemical activity tailoring of a catalyst have been discussed thoroughly. In addition, the practical application of photocatalyst in wastewater treatment is discussed. Further, this chapter also included basic information on the photo-catalytic mechanism and development of photocatalytic systems to the broad readership.

Keywords: Photocatalysts, wastewater treatment, metal sulfides, nanomaterials

3.1 Introduction

Water is, indeed, considered as the most important component for the existence of life on our planet [1]. All known forms of life on Earth require water to survive, and water plays a crucial role in many biological processes, including metabolism, transportation of nutrients, and regulation of body temperature. However, while water is a plentiful resource on our planet, not all of it is readily available for human consumption. Freshwater, which is the type of water that is suitable for human use, is actually a relatively scarce resource. Only about ~3% of the world's water is fresh water, and much of that is locked up in glaciers, polar ice caps, and underground aquifers [2]. As a result, the demand for fresh water to sustain normal life is indeed extremely high, and it is estimated that around 1.2 billion people worldwide lack access to safe drinking water in coming years. The issue of water scarcity is further compounded by factors such as climate change, population growth, and pollution, which can all have a significant impact on the availability and quality of freshwater resources. According to the World Health Organization's latest data, as of 2021, approximately 785 million people worldwide do not have access to basic drinking water services [3]. This means that they are forced to rely on unsafe sources of water, such as surface water, unprotected wells or springs, or water that has been contaminated by human or animal waste. Lack of access to clean drinking water can have serious health consequences, such as the spread of waterborne diseases like cholera, typhoid, and dysentery. It can also contribute to poverty and hinder economic development, as people spend more time and resources collecting and treating water or suffer from illness and reduced productivity.

Recycling and reusing contaminated water can be an important strategy for increasing access to safe water in areas where fresh water resources are scarce or polluted [4]. By treating and reusing water that would otherwise be discarded, we can reduce the pressure on limited water supplies and reduce the amount of untreated wastewater that enters the environment. There are a variety of technologies and methods available for treating contaminated water, such as filtration, disinfection, and reverse osmosis. These processes can remove harmful pollutants and pathogens from the water, making it safe for reuse. Recycling and reusing contaminated water can also have important environmental benefits, such as reducing the amount of water that needs to be withdrawn from rivers and aquifers and reducing the amount of wastewater that is discharged into rivers and oceans [5]. However, it is important to note that not all contaminated water can be safely recycled and reused. Some pollutants, such as heavy metals and certain chemicals, can be difficult to remove through conventional water treatment methods. Additionally, proper management and monitoring are essential to ensure that recycled water is used appropriately and does not pose a risk to public health or the environment.

Photocatalytic wastewater purification with semiconductor photocatalysts is a well-established technique for treating contaminated water [5]. Figure 3.1 shows a representative schematic of semiconductor-based photocatalysis. This process involves the use of a photocatalyst, typically made of a semiconductor material such as metal oxide and metal sulfides, which can be activated by light to generate reactive oxygen species (ROS) that can degrade and remove pollutants from the water [6]. When a photocatalyst is illuminated with light of a specific wavelength, it absorbs the energy and

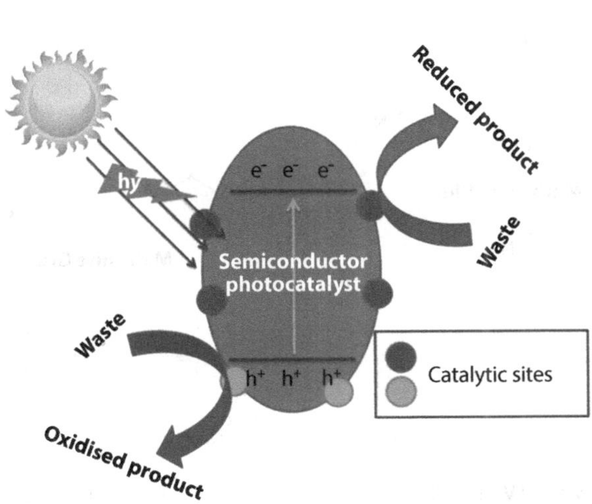

Figure 3.1 Schematic of waste degradation on a metal sulfide surface.

generates electron–hole pairs. These pairs of charged particles react with water and oxygen to form ROS such as hydroxyl radicals (•OH), which can break down organic pollutants into simpler, less harmful compounds. Photocatalytic wastewater treatment has several advantages over traditional treatment methods. It avoids use of any external harmful chemicals, which does not generate additional waste. It can also be effective in treating a wide range of pollutants, including organic compounds, heavy metals, and even some bacteria and viruses. Chemical structures of some dye molecules are presented in Figure 3.2.

However, there are some challenges associated with photocatalytic wastewater treatment. One major challenge is that the process requires a specific wavelength of light to activate the photocatalyst, which depends on the bandgap of the photocatalyst. A wide bandgap photocatalyst required UV light for photocatalytic degradation of pollutants [7]. This can be expensive and energy-intensive and may not be feasible in all settings. Overall, photocatalytic wastewater purification with semiconductor photocatalysts is a promising technology that has the potential to provide safe and clean water to communities around the world. However, further research and development is needed to optimize the process and make it more cost-effective and accessible to communities with limited resources.

A variety of materials have been developed and tested as catalysts for water purification, including metal oxides, metal sulfides, metal–organic frameworks (MOFs), polymers and composites, and other materials. Metal

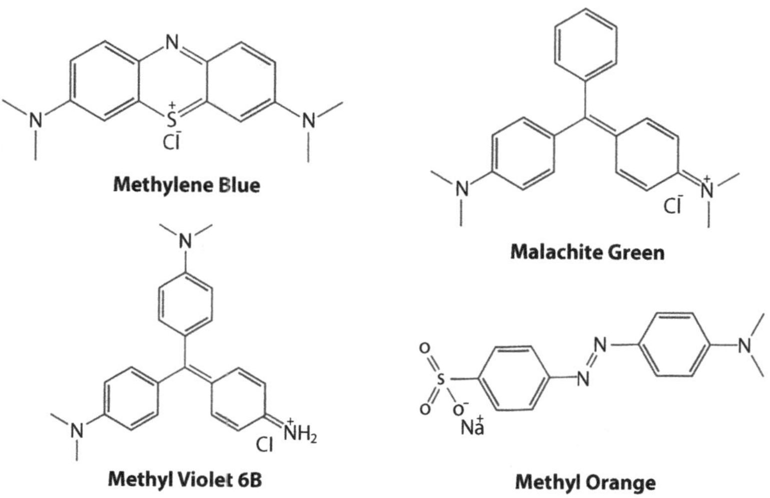

Figure 3.2 Chemical structure of some dyes found in waste water.

oxides such as titanium dioxide (TiO_2), zinc oxide (ZnO), and iron oxide (Fe_2O_3) are some of the most commonly used catalysts for photocatalytic water purification [8]. These materials have high stability and can be easily synthesized in the laboratory, making them attractive for practical applications. MOFs are a newer class of materials that have shown great promise as catalysts for water purification [9]. MOFs are porous materials that can be tuned to selectively adsorb and degrade specific pollutants, making them highly selective and effective for treating contaminated water. Polymers and composites are also being developed as catalysts for water purification. These materials have unique properties that can enhance their catalytic activity and selectivity and can be designed to specifically target certain pollutants. Most of the metal oxides exhibits high bandgap, which is inappropriate to utilize visible light for photocatalysis. Further, the poor environmental stability of MOF restricts their application in real field application in photocatalysis. Metal sulfides, such as cadmium sulfide (CdS), zinc sulfide (ZnS), and copper sulfide (CuS), are widely investigated as photocatalysts for water purification [10]. These materials have high quantum efficiency and can absorb a broader range of light, making them more efficient in certain applications. The notable photocatalytic activity of metal sulfides makes them suitable candidate for solar energy harvesting, water-to-hydrogen production, photocatalytic water decontamination, etc. [11–13]. Overall, the development of new catalysts and materials for water purification is an active area of research and holds great promise for addressing the global water crisis. However, it is important to carefully evaluate the effectiveness and safety of these materials before they are implemented in practical applications.

Metal sulfide–based photocatalytic waste removal from contaminated water has drawn a lot of interest because of its excellent photoabsorption capacity, capacity to use sunlight, and fascinating optoelctronic characteristic [10]. Controlling structural, optical, and electrical features allows for the development of efficient metal sulfide photocatalysts [13]. In this context, nanostructure production is brought to the forefront of metal sulfide photocatalyst design. Many fascinating features of nanostructure metal sulfides contribute in photocatalytic water remediation. The shape and size of nanostructure photocatalysts determine the physiochemical characteristics, which govern the catalytic efficiency [13].

Furthermore, control over crystal structure and exposed facets can lead to improved photocatalytic activity. Efforts are being made to create an efficient metal sulfide nanostructure photocatalyst that can work in visible light. This helps in the use of abundant sunlight as a source of photons. The current chapter attempts to elaborate on the principles of photocatalytic waste removal principle on metal sulfide photocatalyst and degradation mechanism. Additionally, we discuss many metal sulfide photocatalysts that have recently been used for water purification.

3.2 General Working Principle and Important Parameters for Photocatalysis on Metal Sulfide

Photocatalysis on metal sulfides involves the use of semiconducting materials to initiate chemical reactions using light energy. The general working principle of photocatalysis on metal sulfides can be explained in the following steps [14]:

a. **Absorption of light energy:** When metal sulfide is exposed to light, the photons are absorbed by the semiconductor material and promote electrons from the valence band (VB) to the conduction band.
b. **Generation of electron–hole pairs:** This absorption of photons creates electron–hole pairs in the metal sulfide, which can migrate to the surface of the material.
c. **Surface reactions:** The electrons and holes at the surface of the metal sulfide can participate in surface reactions, such as the reduction of oxygen or the oxidation of water, leading to the formation of reactive species such as superoxide or hydroxyl radicals.
d. **Catalytic reactions:** These reactive species can then initiate catalytic reactions with organic pollutants, resulting in the degradation of the pollutant to harmless byproducts.

Important parameters that determine the photocatalytic activity of metal sulfides include the following:

a. **Bandgap energy:** The bandgap energy of a semiconductor material determines the energy of the photons that can be absorbed by the material. A narrow bandgap energy is desirable for photocatalysis, as it allows the material to absorb a wider range of visible light.
b. **Surface area:** A larger surface area of the metal sulfide material allows for more reactive sites for surface reactions, leading to higher catalytic activity.

c. **Crystal structure:** The crystal structure of metal sulfide can affect the reactivity of the material, with certain crystal planes being more active for photocatalytic reactions.
d. **Dopants:** The addition of dopants to the metal sulfide can improve its photocatalytic activity by modifying the bandgap energy or surface properties.
e. **pH:** The pH of the solution can affect the surface charge of the metal sulfide, which can, in turn, affect the adsorption and reaction of pollutants on the material.

Metal sulfide bandgaps are frequently observed to be less than those of their oxide counterparts. It is generally true that metal sulfide bandgaps are less than those of their oxide counterparts. This is because sulfide ions (S^{2-}) have a larger ionic radius compared to oxide ions (O^{2-}), which results in a wider separation between the valence and conduction bands in sulfide semiconductors [14]. The wider bandgap energy of oxide semiconductors means that they can only absorb higher energy photons in the ultraviolet (UV) range, whereas sulfide semiconductors can absorb a wider range of visible light, making them more suitable for solar-assisted photocatalytic applications [15]. Moreover, metal sulfides tend to have higher absorption coefficients than their oxide counterparts, meaning that they can absorb more photons of the same energy, which can result in higher photocatalytic activity. Additionally, metal sulfides can exhibit narrower bandgap energies than their oxide counterparts due to the formation of impurity states or defects in the material, which can improve the efficiency of photocatalysis by facilitating charge separation and reducing recombination of electron–hole pairs. However, it should be noted that the choice of semiconductor material for a particular photocatalytic application depends on various factors, including the type of pollutant, reaction conditions, and the availability and cost of the material. Hence, a comparative study of different semiconductors should be performed to identify the most suitable material for a specific photocatalytic application.

3.3 Different Types of Metal Sulfide Photocatalysts: Monometallic Sulfides, Multimetallic Metal Sulfides, and Metal Sulfide Heterostructure

Metal sulfide photocatalysts can be broadly categorized into three types: monometallic sulfides, multimetallic sulfides, and metal sulfide heterostructures. Table 3.1 shows the list of metal sulfides used in photocatalysis.

3.3.1 Monometallic Sulfides

These are single-metal sulfide photocatalysts, such as cadmium sulfide (CdS), zinc sulfide (ZnS), and copper sulfide (CuS). Monometallic sulfides have high photocatalytic activity, good chemical stability, and low cost. They are widely used in various photocatalytic applications, such as water splitting, pollutant degradation, and hydrogen production [16].

3.3.2 Multimetallic Sulfides

These are sulfide photocatalysts composed of two or more metals, such as cobalt sulfide (CoS), nickel sulfide (NiS), and iron sulfide (FeS). Multimetallic sulfides exhibit higher photocatalytic activity than monometallic sulfides due to their unique electronic and structural properties. They are used in various photocatalytic applications, such as CO_2 reduction and water splitting [17].

3.3.3 Metal Sulfide Heterostructures

These are photocatalysts composed of two or more different metal sulfides, such as CdS/ZnS, CdS/CuS, and CdS/FeS. Metal sulfide heterostructures have a high surface area and unique electronic properties, which make them highly efficient photocatalysts [18]. They are widely used in various photocatalytic applications, such as water splitting and pollutant degradation. Some of the metals sulfides used in photocatalysis is listed in the following table (Table 3.1).

In the following section, we will discuss recent reports on photocatalytic waste water treatment using some specific metal sulfides.

3.4 Copper Sulfide as Photocatalyst for Water Remediation

Copper sulfide (CuS) has been studied as a potential photocatalyst for water remediation due to its unique optical, electronic, and chemical properties [19]. CuS can be used for the degradation of organic pollutants, reduction of heavy metals, and removal of bacteria and viruses from water. It has a narrow bandgap semiconductor that can absorb visible light, which makes it an attractive option for photocatalytic applications. When exposed to light, CuS can generate electron–hole pairs, which can then participate in

Table 3.1 Metal sulfides used in photocatalysis.

Mono-metal sulfides					
Ce_2S_3	Rh_2S_3	α-MnS	CoS	CuS	OsS_2
α-In_2S_3	PdS	β-MnS	CoS_2	CuS_2	CrS
β-In_2S_3	PdS_2	γ-MnS	Co_2S_3	$Cu_{1.12}S$	Cr_2S_3
γ-In_2S_3	ReS_2	FeS	Co_3S_4	$Cu_{1.39}S$	GeS
TiS_2	BaS	FeS_2	Co_4S_3	$Cu_{1.6}S$	Ni_3S_4
$Ti_{0.67}S$	PtS	Fe_3S_4	Co_9S_8	$Cu_{1.75}S$	$Ni_{3+x}S_2$
PbS	Sb_2S_3	WS_2	ZnS	$Cu_{1.8}S$	Ni_3S_2
TaS_2	Sn_4S_5	IrS_2	CdS	$Cu_{1.96}S$	NiS_2
ZrS	Sn_3S_4	Ir_2S_3	CaS	Cu_2S	Ni_6S_5
TeS_2	Sn_2S_3	NbS	Li_2S	Ag_2S	Ni_7S_6
VS_2	SnS_2	NbS_2	GeS_2	MoS_2	Ni_9S_8
In_2S_3	SnS	GeS	Bi_2S_3	MoS_3	NiS
YbS	BeS	Tl_2S	Rb_2S_3	Cs_2S_5	SmS

redox reactions. The resulting reactive species, such as hydroxyl radicals, can then degrade organic pollutants and kill bacteria and viruses. Several methods have been developed for the synthesis of CuS photocatalysts, including chemical precipitation, hydrothermal synthesis, solvothermal synthesis, and microwave-assisted synthesis [19, 20]. The choice of synthesis method can influence the properties of the CuS photocatalyst, such as particle size, morphology, and crystal structure.

CuS_x nanostructures, including CuS and CuS_2, have been studied as potential photocatalysts for water remediation due to their unique properties [21]. CuS_x nanostructures having surface area allows more efficient absorption of light and increased surface reactions. They also have tunable bandgap energies, which can be adjusted by changing the stoichiometry of Cu and S, making them suitable for visible light photocatalysis. CuS_x nanostructures can be synthesized using various methods, such as solvothermal synthesis, hydrothermal synthesis, and chemical vapor deposition. These methods can produce different nanostructures, such as nanoparticles, nanorods, and nanosheets, with different sizes and morphologies. The properties of CuS_x

nanostructures can be further tuned by controlling the synthesis parameters, such as temperature, pressure, and precursor concentrations. Several studies have shown that CuS_x nanostructures can effectively degrade organic pollutants, such as dyes and phenols, under visible light irradiation. For example, CuS nanoparticles were used to degrade methyl orange, and CuS_2 nanoparticles were used to degrade Rhodamine B [21]. CuS_x nanostructures have also been used for the removal of heavy metals from water, such as Cu(II) and Cr(VI), through photocatalytic reduction [22]. The photocatalytic activity of CuS_x nanostructures can be enhanced by doping with other elements, such as nitrogen, to increase their visible light absorption and surface reactions. Additionally, CuS_x nanostructures can be combined with other photocatalysts, such as TiO_2, to form composite materials with synergistic effects [23]. However, the practical application of CuS_x nanostructures for water remediation still faces several challenges, such as low stability and reproducibility, difficulty in separation and recovery, and potential environmental risks. Further research is needed to optimize the synthesis and properties of CuS_x nanostructures and to evaluate their long-term stability and potential impacts on the environment.

There have been several studies on the synthesis of CuS_x nanostructures, including CuS and Cu_5S_9, using various methods. Li *et al.* reported a simple hydrothermal method for the synthesis of CuS and Cu_5S_9 nanostructures [24]. In their study, they used copper acetate and thiourea as the precursors and a mixture of water and ethylene glycol as the solvent. The reaction was carried out at 200°C for 8 h, and the resulting products were characterized by various techniques, such as X-ray diffraction, scanning electron microscopy, and transmission electron microscopy. The study showed that the size and morphology of the CuS and Cu_5S_9 nanostructures could be controlled by adjusting the reaction conditions, such as precursor concentration, reaction temperature, and reaction time. The CuS nanostructures synthesized by Li *et al.* had a size range of 20–50 nm and a morphology of nanorods, whereas the Cu_5S_9 nanostructures had a size range of 30–100 nm and a morphology of irregular particles. The CuS and Cu_5S_9 nanostructures synthesized by Li *et al.* were also evaluated for their photocatalytic activity. The results showed that the Cu_5S_9 nanostructures had a higher photocatalytic activity than the CuS nanostructures for the degradation of Rhodamine B under visible light irradiation. Figure 3.3 presents a mechanism of Rhodamine B degradation on CuS photocatalyst. The enhanced photocatalytic activity of Cu_5S_9 was attributed to its narrow bandgap energy and the formation of a heterojunction between the CuS and Cu_2S phases [25].

Chlorotris(triphenylphosphine)-copper(I) ($[CuCl(PPh_3)]$) and Cu_4(tu) 94•$4H_2O$ (where tu is thiourea) have been used as precursors to synthesize

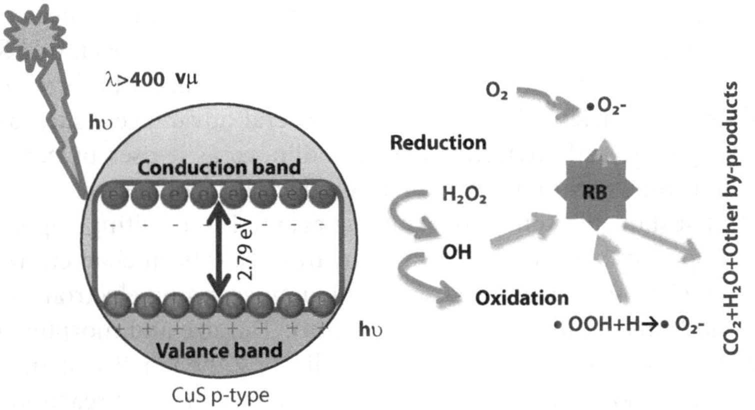

Figure 3.3 Schematic of photocatalytic Rhodamine B degradation on p-type CuS semiconductor.

Cu_2S and CuS, respectively [26]. [$CuCl(PPh_3)$] has been used as a precursor to synthesize Cu_2S nanostructures by a solvothermal method. In this method, [$CuCl(PPh_3)$] is dissolved in a solvent, such as ethylene glycol or dimethylformamide, along with a sulfur source, such as thiourea or sulfur powder. The mixture is then heated at a high temperature under pressure to induce the formation of Cu_2S nanostructures. The size and morphology of the Cu_2S nanostructures can be controlled by adjusting the reaction conditions, such as the precursor concentration, reaction temperature, and reaction time. Similarly, $Cu_4(tu)94•4H_2O$ has been used as a precursor to synthesize CuS nanoparticles by a simple precipitation method. In this method, $Cu_4(tu)_{94}•4H_2O$ is dissolved in water, and a sulfur source, such as thioacetamide, is added to the solution. The mixture is then heated and stirred for a certain period of time to form CuS nanoparticles. The size and morphology of the CuS nanoparticles can be controlled by adjusting the reaction conditions, such as the precursor concentration, pH, and reaction time. Both of these precursors have been successfully used to synthesize Cu_2S and CuS nanostructures with tunable sizes and morphologies. These nanostructures have been studied for their photocatalytic activity for the degradation of various organic pollutants and have shown promising results. However, further research is still needed to optimize the synthesis and properties of these nanostructures for practical applications in water remediation.

There have been studies reporting the synthesis of copper sulfide nanoparticles using a simple one-step method based on biogenic H_2S generated from water remediation [27]. In this method, the copper precursor, such as copper acetate or copper nitrate, is mixed with biogenic

H_2S generated from water remediation, and the mixture is heated at a certain temperature for a specific period of time. The biogenic H_2S acts as a reducing agent and reacts with the copper precursor to form copper sulfide nanoparticles. This method has several advantages, such as low cost, simplicity, and environmental friendliness, as it uses biogenic H_2S generated from water remediation as a reducing agent, which is a renewable and sustainable source of reducing agent. The resulting copper sulfide nanoparticles synthesized by this method have been characterized by various techniques, such as X-ray diffraction, scanning electron microscopy, and transmission electron microscopy. The size and morphology of the nanoparticles can be controlled by adjusting the reaction conditions, such as precursor concentration, reaction temperature, and reaction time. The synthesized copper sulfide nanoparticles have been evaluated for their photocatalytic activity for the degradation of various organic pollutants in water. The results showed that the copper sulfide nanoparticles have high photocatalytic activity under visible light irradiation, which can be attributed to their narrow bandgap energy and large specific surface area. The simple one-step method based on biogenic H_2S generated from water remediation provides a promising approach for the synthesis of copper sulfide nanoparticles with potential applications in water remediation.

In recent years, there have been several studies on the use of CuS photocatalysts for water remediation. For example, CuS nanoparticles were used for the degradation of the antibiotic tetracycline in water, and CuS nanorods were used for the removal of Cr(VI) from water [28]. CuS has also been used in combination with other photocatalysts, such as TiO_2, to enhance its photocatalytic activity. While CuS shows promise as a photocatalyst for water remediation, more research is needed to optimize its properties and develop efficient and scalable synthesis methods. Additionally, the long-term stability and potential environmental impacts of CuS photocatalysts should be carefully evaluated before their widespread use in water remediation applications.

3.5 ZnS as a Photocatalyst for Water Treatment

ZnS has been extensively studied as a photocatalyst for water treatment due to its unique properties, such as a suitable bandgap energy, high photoactivity, and good stability. ZnS can be synthesized in various forms, including nanoparticles, nanosheets, nanorods, and hollow spheres, using different synthetic methods such as solvothermal, hydrothermal, and sonochemical methods [29]. The size and morphology of the ZnS nanoparticles can be tuned by adjusting the reaction conditions. ZnS photocatalysts have been shown to have high photocatalytic activity in the degradation of various pollutants, including dyes, pesticides, and organic compounds. The photocatalytic activity of ZnS photocatalysts can be enhanced by doping with other metals or non-metals, such as Ag, Au, Cu, S, and N, or by coupling with other semiconductors, such as TiO_2 and CdS.

When ZnS is exposed to light, photons with energy higher than the bandgap energy of ZnS can be absorbed by the semiconductor, promoting an electron from the VB to the conduction band, leaving behind a hole in the VB. These electron–hole pairs can migrate to the surface of the photocatalyst, where they can react with adsorbed species, such as water or organic pollutants, to generate radicals that can break down the pollutants into harmless products. The photogenerated electrons can reduce adsorbed oxygen molecules to form superoxide radicals ($O_2\bullet^-$), whereas the holes can oxidize water molecules to form hydroxyl radicals (•OH). These radicals can then react with organic pollutants adsorbed on the surface of the photocatalyst, leading to their degradation into simpler and less harmful compounds. The resulting products can then be mineralized into carbon dioxide, water, and other harmless species. The efficiency of the photocatalytic activity of ZnS can be influenced by various factors, such as the size and morphology of the ZnS nanoparticles, the intensity and wavelength of the light source, the pH of the solution, and the concentration and type of the pollutants. Therefore, optimizing these factors is essential to enhance the photocatalytic efficiency of ZnS for water treatment applications.

In the photocatalytic degradation of methylene blue (MB) using a semiconductor photocatalyst, the primary oxidant responsible for the degradation of MB is the hydroxyl radical (•OH). When the photocatalyst is irradiated with light, the VB electrons of the semiconductor are excited to the conduction band (CB), leaving behind holes (h+) in the VB. The photogenerated electrons and holes can then participate in redox reactions with adsorbed species, such as oxygen or water, to generate reactive species, including hydroxyl radicals (•OH), superoxide radicals ($O2\bullet^-$), and hydrogen peroxide (H_2O_2), among others. Among these reactive species, hydroxyl radicals (•OH) are the most potent oxidants and are primarily responsible for the degradation of MB. The hydroxyl radicals can attack the MB molecule by abstracting hydrogen atoms from the benzene rings or breaking the carbon-nitrogen bond, leading to the formation of simpler and less harmful products. The degradation of MB can be monitored by measuring the decrease in the absorbance of the solution at the characteristic wavelength of MB. The formation of hydroxyl radicals (•OH) in the photocatalytic degradation of MB can be attributed to the high oxidizing power of the holes (h+) generated by the photocatalyst. The holes can react with adsorbed water molecules to form hydroxyl radicals (•OH) by a process called the "Hole Oxidation Reaction" (HOR):

$$h^+ + H2O \rightarrow \bullet OH + H^+$$

The hydroxyl radicals (•OH) can then react with the adsorbed MB molecule to initiate the degradation process. In summary, the photocatalytic degradation of MB involves the generation of hydroxyl radicals (•OH) as the primary oxidant, which are responsible for the degradation of MB into simpler and less harmful products. Figure 3.4 presents a mechanism involve in the photocatalytic degradation of MB on ZnS.

Chitosan-g-poly acrylamide/ZnS (ChPA/ZS) nanocomposite has been found to exhibit good photocatalytic activity for the degradation of Congo red and Methyl orange dyes [30]. The ChPA/ZS nanocomposite is synthesized by graft copolymerization of acrylamide onto chitosan followed by the deposition of ZnS nanoparticles onto the ChPA backbone. The resulting nanocomposite exhibits good stability, biocompatibility, and low toxicity. The photocatalytic activity of the ChPA/ZS nanocomposite is attributed to the synergistic effect of chitosan, polyacrylamide, and ZnS nanoparticles. Chitosan provides a porous and hydrophilic matrix for the immobilization of ZnS nanoparticles, whereas polyacrylamide offers excellent photostability and electron transfer properties. ZnS nanoparticles act as a

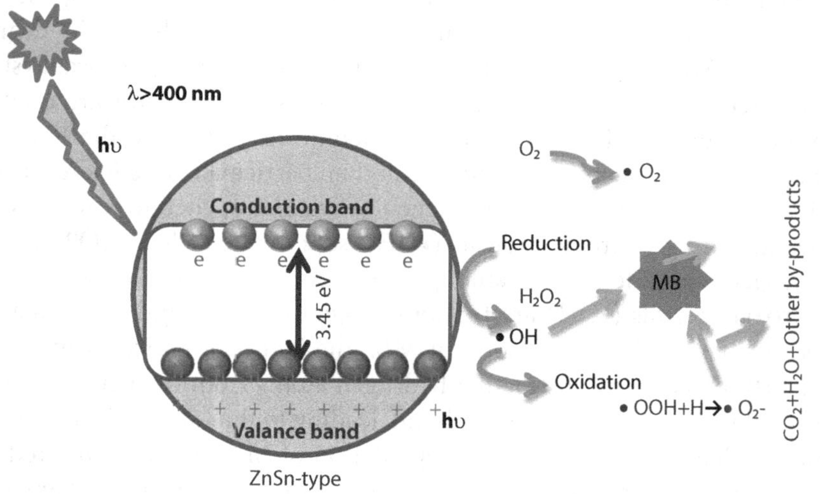

Figure 3.4 Schematic of photocatalytic methylene blue degradation on n-type ZnS semiconductor.

semiconductor and promote the separation of electron–hole pairs under UV irradiation. The photocatalytic degradation of Congo red and Methyl orange dyes by the ChPA/ZS nanocomposite was studied under UV light irradiation. The results showed that the ChPA/ZS nanocomposite exhibited excellent photocatalytic activity for the degradation of both dyes. The degradation rate of Congo red was found to be 98.6% within 150 min of irradiation, whereas the degradation rate of Methyl orange was found to be 92.3% within 180 min of irradiation. In conclusion, the ChPA/ZS nanocomposite has been found to be an efficient photocatalyst for the degradation of Congo red and Methyl orange dyes. The ChPA/ZS nanocomposite has potential applications in wastewater treatment, environmental remediation, and other areas where photocatalysis is required.

RGO-ZnS composites are materials made up of reduced graphene oxide (RGO) and zinc sulfide (ZnS) particles that are combined to form a hybrid structure. These composites have been shown to have excellent photocatalytic properties, meaning that they can use light energy to catalyze chemical reactions. Rhodamine B is a dye molecule that is commonly used as a test molecule for studying photocatalytic activity [31]. When exposed to light in the presence of a photocatalyst such as RGO-ZnS composites, Rhodamine B can be degraded into smaller, less harmful molecules. Studies have shown that RGO-ZnS composites are effective photocatalysts for the degradation of Rhodamine B. The presence of graphene oxide in the composite structure can enhance the photocatalytic activity of ZnS, leading to faster degradation of Rhodamine B. Overall, the combination of RGO and ZnS in a composite structure has shown promise for the development of highly efficient and effective photocatalysts for a range of environmental and industrial applications.

Zinc sulfide (ZnS) and iron-doped ZnS are semiconductor materials that have been investigated for their potential to reduce hexavalent chromium [Cr(VI)] in water [32]. Hexavalent chromium is a toxic and carcinogenic pollutant that can be found in industrial wastewaters and poses a serious threat to human health and the environment. When exposed to light, ZnS and iron-doped ZnS can generate electron–hole pairs, which can then react with water molecules to form highly ROS. These reactive species can reduce Cr(VI) to less toxic and less mobile trivalent chromium [Cr(III)]. Studies have shown that both ZnS and iron-doped ZnS can effectively reduce Cr(VI) in water under UV or visible light irradiation. Iron-doped ZnS has been shown to be more effective than pure ZnS due to the presence of iron, which acts as a co-catalyst and enhances the electron transfer process. Overall, the use of ZnS and iron-doped ZnS as photocatalysts for the reduction of Cr(VI) in water shows great promise for the development of effective and sustainable water treatment technologies.

ZnS quantum dots (QDs), as well as gold (Au)/ZnS and silver (Ag)/ZnS nanocomposites, have been studied for their photocatalytic activity in the degradation of MB, a common organic dye pollutant [33]. ZnS QDs are semiconductor materials that can generate electron–hole pairs when exposed to light. These electron–hole pairs can then react with water molecules to produce highly ROS, which can oxidize and degrade organic pollutants like MB. Au/ZnS and Ag/ZnS nanocomposites are hybrid materials that combine the photocatalytic activity of ZnS with the plasmonic properties of gold or silver nanoparticles. These nanoparticles can enhance the absorption of light and the generation of electron–hole pairs, leading to improved photocatalytic activity. In studies evaluating the photocatalytic activity of these materials, it has been found that ZnS QDs, Au/ZnS, and Ag/ZnS all show good activity in the degradation of MB under visible light irradiation. Au/ZnS and Ag/ZnS nanocomposites have been shown to be more effective than pure ZnS QDs due to the plasmonic enhancement effect. Overall, ZnS is a promising photocatalyst for water treatment due to its high photoactivity, low cost, and environmental friendliness. However, further research is needed to optimize the synthesis and properties of ZnS photocatalysts for practical applications in water treatment.

3.6 FeS as a Photocatalyst for Water Remediation

Iron sulfides have been investigated for their potential use as photocatalysts in environmental applications due to their unique properties. Pyrite (FeS_2) is a common iron sulfide mineral that has been studied for its photocatalytic properties [34]. It has a direct bandgap of approximately 0.95 eV, making it suitable for absorbing visible light, and has a high absorption coefficient, as you noted. Pyrite has also been found to be stable under visible light irradiation, which is important for photocatalytic applications. Pyrrhotite ($Fe_{1-x}S$) is another iron sulfide mineral that has been investigated for its photocatalytic properties [35]. It has a more complex crystal structure than pyrite and can have varying compositions. Pyrrhotite has been found to have a bandgap energy ranging from 0.8 eV to 1.2 eV, which makes it suitable for absorbing visible light. Troilite (FeS) is a simple iron sulfide mineral that has been used in a variety of applications, including as a photocatalyst. Troilite has a direct bandgap of approximately 1.3 eV, which makes it suitable for absorbing visible light. Generally, the unique properties of iron sulfides make them a promising material for use as photocatalysts in environmental applications. However, more research is needed to fully understand their potential and optimize their properties for specific applications.

FeS/ZnO nanoarrays have been investigated for their piezophototronic and photocatalytic coupling effect, which is the synergistic effect that occurs when the piezoelectric and photovoltaic effects of the material are coupled together [36]. Piezophototronic effect refers to the phenomenon where the mechanical stress or strain applied to a piezoelectric material can alter its electrical and optical properties, resulting in changes in its photocatalytic activity. ZnO is a well-known piezoelectric material, which can produce piezoelectric charges upon the application of mechanical stress or strain. FeS, on the other hand, is a good photocatalyst that can absorb visible light and has high electron mobility. FeS/ZnO nanoarrays have been shown to exhibit enhanced activity due to the piezophototronic effect. When an external mechanical force is applied to the FeS/ZnO nanoarrays, it induces a piezoelectric charge in the ZnO, which, in turn, alters the band structure of FeS and enhances its photocatalytic activity. In addition, FeS/ZnO nanoarrays have also been shown to exhibit enhanced photovoltaic properties due to the piezophototronic effect. The piezoelectric charges generated in the ZnO under mechanical stress or strain can be used to drive a photocurrent in the FeS/ZnO nanoarrays. The piezophototronic/photocatalytic coupling effect of FeS/ZnO nanoarrays holds great potential for various applications, including in energy harvesting, environmental remediation, and sensing. However, further research is needed to optimize the properties and performance of these materials for secific applications.

FeS/ZnO nanoarrays have been shown to exhibit significant sono-photocatalytic activity for decomposing MB in water under ultrasonic and solar irradiation [37]. Sono-photocatalysis is a combination of sonocatalysis and photocatalysis, which involves the use of both ultrasonic waves and solar irradiation to activate the photocatalytic process. In this process, the ultrasonic waves can enhance the mass transfer and mixing of reactants while also generating cavitation bubbles, which can further enhance the photocatalytic activity of the material. FeS/ZnO nanoarrays have been found to be efficient sono-photocatalysts for decomposing MB in water under ultrasonic and solar irradiation. The nanoarrays exhibit high catalytic activity due to their unique structural and electronic properties, which allow for efficient electron–hole separation and charge transfer during the photocatalytic process. The sono-photocatalytic activity of FeS/ZnO nanoarrays has been attributed to the synergistic effect between ultrasonic waves and solar irradiation, which can enhance the photocatalytic efficiency of the material. This effect is due to the combination of mechanical and photonic energies, which can promote the generation of reactive species and enhance the degradation of the target pollutant. The sono-photocatalytic activity of FeS/ZnO nanoarrays holds great potential for various applications in environmental remediation and wastewater treatment. However, further research is needed to optimize the properties and performance of these materials for specific applications.

Fe_2O_3/FeS nanocomposites have been investigated as a promising photocatalyst for water treatment applications [38]. These nanocomposites are composed of iron oxide (Fe_2O_3) and iron sulfide (FeS) nanoparticles, which are known to exhibit excellent photocatalytic properties. Iron oxide nanoparticles are well-known photocatalysts that can absorb visible light and generate electron–hole pairs, which can then promote the degradation of organic pollutants in water. However, iron oxide has limited photocatalytic activity in the visible light range and can be prone to charge carrier recombination, which can limit its efficiency. Iron sulfide, on the other hand, is a good visible-light-active photocatalyst that can enhance the photocatalytic activity of iron oxide nanoparticles. Iron sulfide can absorb visible light due to its narrow bandgap and can generate electrons and holes that can be utilized for photocatalytic reactions. Fe_2O_3/FeS nanocomposites have been found to exhibit improved photocatalytic activity for water treatment applications compared to individual iron oxide or iron sulfide nanoparticles. The nanocomposites have a synergistic effect, where the iron sulfide nanoparticles can act as co-catalysts, facilitating the separation of photo-generated electrons and holes and reducing the charge carrier recombination rate in the iron oxide nanoparticles. In addition, Fe_2O_3/FeS nanocomposites have been shown to exhibit good stability and reusability, making them promising for practical applications in water treatment. Fe_2O_3/FeS nanocomposites hold great potential as a photocatalyst for water treatment applications.

The TiO_2/FeS_2 system has been extensively investigated as a potential solar photocatalyst for various environmental applications such as water treatment and air purification [39]. TiO_2 is a well-known photocatalyst that can absorb UV light and generate electron–hole pairs, which can be utilized for photocatalytic reactions. However, TiO_2 has limited photocatalytic activity in the visible light range, which limits its efficiency. Iron disulfide (FeS_2) is a promising visible light-active photocatalyst that can absorb visible light due to its narrow bandgap and generate electrons and holes that can be used for photocatalytic reactions. FeS_2 has excellent optical and electronic properties and is abundant and low-cost, making it a promising material for solar photocatalysis applications.

The combination of TiO_2 and FeS_2 in a photocatalytic system can provide a wider range of light absorption and improved photocatalytic activity. The FeS_2 nanoparticles can act as co-catalysts, facilitating the separation of photo-generated electrons and holes in the TiO_2 nanoparticles and reducing the charge carrier recombination rate. This enhances the overall photocatalytic efficiency of the system. Recent studies have demonstrated that the TiO_2/FeS_2 system exhibits excellent photocatalytic activity for various applications, such as the degradation of organic pollutants in water and the removal of air pollutants. The system has also been found to be stable and reusable, making it a promising candidate for practical applications in solar photocatalysis. In summary, the TiO_2/FeS_2 system is a promising solar photocatalyst for environmental applications due to its excellent photocatalytic activity and stability.

Pyrite (FeS_2) photocatalysis has been investigated as a potential method for the synergistic removal of heavy metal Cr(VI) and organic matter malachite green from water [40]. Malachite green is a synthetic organic dye commonly used in the textile industry and can have harmful effects on the environment and human health if not properly removed from wastewater. Cr(VI) is a highly toxic heavy metal that can cause serious health problems even at low concentrations. Under UV irradiation, pyrite nanoparticles can generate ROS such as hydroxyl radicals, which can oxidize and degrade both malachite green and Cr(VI) in water. Pyrite has a narrow bandgap that allows it to absorb visible and UV light, which makes it an effective photocatalyst under both solar and UV irradiation. Recent studies have shown that the pyrite photocatalysis method can effectively remove both malachite green and Cr(VI) from water with high efficiency. The removal rates of malachite green and Cr(VI) depend on several factors such as the concentration of the contaminants, the pH of the solution, and the irradiation intensity. For example, one study reported that, under UV irradiation, pyrite nanoparticles could remove up to 94.4% of Cr(VI) and 99.7% of malachite green from contaminated water within 60 min. The study also found that the removal efficiency was higher under acidic conditions and increased with increasing irradiation intensity. Pyrite photocatalysis holds great potential as a method for the synergistic removal of heavy metals and organic contaminants from water. Further research is needed to optimize the properties and performance of the pyrite nanoparticles for specific applications and to better understand the mechanism of action of pyrite photocatalysis.

3.7 Conclusion

Metal sulfide semiconductor (MSS) water treatment technology has a promising future in the field of water treatment. MSS technology utilizes semiconductor materials such as titanium dioxide (CuS, ZnS, etc.) to generate ROS and free radicals that can degrade organic pollutants and kill bacteria and viruses in water. One of the major advantages of MSS technology is that it can work in the visible light, which does not require any chemical additives or the addition of external energy sources such as UV radiation. This makes MSS as a cost-effective and sustainable water treatment option. Additionally, MSS technology has the potential to be implemented in wide range of applications, including the treatment of municipal and industrial wastewater, as well as the purification of drinking water. In the future, MSS technology is likely to see increased adoption and development. Researchers are currently exploring new semiconductor materials and optimizing the design of MSS devices to improve their efficiency and performance. Additionally, the integration of MSS technology with other water treatment methods, such as membrane filtration and activated carbon adsorption, could lead to even more effective and versatile water treatment systems. Overall, the future of MSS water treatment looks bright, and it has the potential to play an important role in addressing global water scarcity and pollution challenges. However, the issues such as photocorrosion, metal leaching, and long-term instability of metal sulfide catalyst have to be keep in mind and addressed properly to make MSS successful.

References

1. Tortajada, C., Contributions of recycled wastewater to clean water and sanitation Sustainable Development Goals. *NPJ Clean Water*, 3, 1, p.22, 2020.
2. Moschell, M. P. Governing the Right to Water. *J. Transnat'l L. & Pol'y*, 29, 199, 2019.
3. https://www.who.int/news-room/fact-sheets/detail/drinking-water.
4. Sarver, B. Why We Should Reuse Wastewater. Certified Public Manager@ Applied Research, 3, 1, 3, 2022.
5. Mills, A. and Le Hunte, S., An overview of semiconductor photocatalysis. *Photochem. Photobiol. A*, *108*, 1, 1–35, 1997.
6. Opoku, F., Govender, K.K., van Sittert, C.G.C.E., Govender, P.P., Recent progress in the development of semiconductor-based photocatalyst materials for applications in photocatalytic water splitting and degradation of pollutants. *Adv. Sustain. Syst.*, *1*, 7, 1700006, 2017.

7. Roy, N., Sohn, Y., Pradhan, D., Synergy of low-energy {101} and high-energy {001} TiO2 crystal facets for enhanced photocatalysis. *ACS Nano*, *7*, 3, 2532–2540, 2013.
8. Kormann, C., Bahnemann, D.W., Hoffmann, M.R., Environmental photochemistry: Is iron oxide (hematite) an active photocatalyst? A comparative study: α-Fe2O3, ZnO, TiO2. *J. Photochem. Photobiol. A*, *48*, 1, 161–169, 1989.
9. Jun, B.M., Al-Hamadani, Y.A., Son, A., Park, C.M., Jang, M., Jang, A., Kim, N.C., Yoon, Y., Applications of metal-organic framework based membranes in water purification: A review. *Sep. Purif. Technol.*, *247*, 116947, 2020.
10. Sahoo, A.K., Srivastava, S.K., Raul, P.K., Gupta, A.K., Shrivastava, R., Graphene nanocomposites of CdS and ZnS in effective water purification. *J. Nanopart. Res.*, *16*, 1–17, 2014.
11. Ghribi, F., Alyamani, A., Ayadi, Z.B., Djessas, K., Mir, L.E., Study of CuS thin films for solar cell applications sputtered from nanoparticles synthesised by hydrothermal route. *Energy Proc.*, *84*, 197–203, 2015.
12. Patra, B.K., Khilari, S., Pradhan, D., Pradhan, N., Hybrid dot–disk Au-CuInS2 nanostructures as active photocathode for efficient evolution of hydrogen from water. *Chem. Mater.*, *28*, 12, 4358–4366, 2016.
13. Kundu, J., Khilari, S., Pradhan, D., Shape-dependent photocatalytic activity of hydrothermally synthesized cadmium sulfide nanostructures. *ACS Appl. Mater. Interfaces*, *9*, 11, 9669–9680, 2017.
14. Zhang, S., Ou, X., Xiang, Q., Carabineiro, S.A., Fan, J., Lv, K., Research progress in metal sulfides for photocatalysis: From activity to stability. *Chemosphere*, 300, 135085, 2022.
15. Patra, B.K., Khilari, S., Bera, A., Mehetor, S.K., Pradhan, D., Pradhan, N., Chemically filled and Au-coupled BiSbS3 nanorod heterostructures for photoelectrocatalysis. *Chem. Mater.*, *29*, 3, 1116–1126, 2017.
16. Ayodhya, D. and Veerabhadram, G., Facile fabrication, characterization and efficient photocatalytic activity of surfactant free ZnS, CdS and CuS nanoparticles. *J. Sci.: Adv. Mater. Devices*, *4*, 3, 381–391, 2019.
17. Guo, Y., Guo, D., Ye, F., Wang, K., Shi, Z., Synthesis of lawn-like NiS2 nanowires on carbon fiber paper as bifunctional electrode for water splitting. *Int. J. Hydrogen Energy*, *42*, 27, 17038–17048, 2017.
18. Muslim, Z.R. and Kadhim, R.F., Photocatalytic removal of methylene blue dye by using of ZnS and CdS. *Iraqi J. Phys.*, *15*, 33, 11–16, 2017.
19. Kundu, J. and Pradhan, D., Influence of precursor concentration, surfactant and temperature on the hydrothermal synthesis of CuS: structural, thermal and optical properties. *New J. Chem.*, *37*, 5, 1470–1478, 2013.
20. Nethravathi, C., Rajamathi, J.T., Rajamathi, M., Microwave-assisted synthesis of porous aggregates of CuS nanoparticles for sunlight photocatalysis. *ACS Omega*, *4*, 3, 4825–4831, 2019.
21. Shamraiz, U., Badshah, A., Hussain, R.A., Nadeem, M.A., Saba, S., Surfactant free fabrication of copper sulphide (CuS–Cu2S) nanoparticles from single

source precursor for photocatalytic applications. *J. Saudi Chem. Soc, 21*, 4, 390–39, 2017.

22. ul Ain, N., ur Rehman, Z., Nayab, U., Nasir, J.A., Aamir, A., Facile photocatalytic reduction of carcinogenic Cr (VI) on Fe-doped copper sulfide nanostructures. *RSC Adv., 10*, 46, 27377–27386, 2020.
23. Khanchandani, S., Kumar, S., Ganguli, A.K., Comparative study of TiO2/CuS core/shell and composite nanostructures for efficient visible light photocatalysis. *ACS Sustain. Chem. Eng., 4*, 3, 1487–149, 2016.
24. Li, S., Zhang, Z., Yan, L., Jiang, S., Zhu, N., Li, J., Li, W., Yu, S., Fast synthesis of CuS and Cu9S5 microcrystal using subcritical and supercritical methanol and their application in photocatalytic degradation of dye in water. *J. Supercrit. Fluids, 123*, 11–17, 2017.
25. Leidinger, P., Popescu, R., Gerthsen, D., Lünsdorf, H., Feldmann, C., Nanoscale copper sulfide hollow spheres with phase-engineered composition: covellite (CuS), digenite (Cu1.8S), chalcocite (Cu2S). *Nanoscale, 3*, 6, 2544–2551, 2011.
26. Kumar, P., Gusain, M., Nagarajan, R., Synthesis of Cu1.8S and CuS from copper-thiourea containing precursors; anionic (Cl–,NO3–,SO42–) influence on the product stoichiometry. *Inorg. Chem., 50*, 7, 3065–3070, 2011.
27. Colipai, C., Southam, G., Oyarzún, P., González, D., Díaz, V., Contreras, B., Nancucheo, I., Synthesis of copper sulfide nanoparticles using biogenic H2S produced by a low-pH sulfidogenic bioreactor. *Minerals, 8*, 2, 35, 2018.
28. Wu, Z., Liu, X., Yu, C., Li, F., Zhou, W., Wei, L., Construct interesting CuS/TiO2 architectures for effective removal of Cr (VI) in simulated wastewater via the strong synergistic adsorption and photocatalytic process. *Sci. Total Environ., 796*, 148941, 2018.
29. Fang, X., Zhai, T., Gautam, U.K., Li, L., Wu, L., Bando, Y., Golberg, D., ZnS nanostructures: from synthesis to applications. *Prog. Mater. Sci., 56*, 2, 175–287, 2011.
30. Pathania, D., Gupta, D., Ala'a, H., Sharma, G., Kumar, A., Naushad, M., Ahamad, T., Alshehri, S.M., Photocatalytic degradation of highly toxic dyes using chitosan-g-poly (acrylamide)/ZnS in presence of solar irradiation. *J. Photochem. Photobiol. A, 329*, 61–68, 2016.
31. Chakraborty, K., Chakrabarty, S., Das, P., Ghosh, S., Pal, T., UV-assisted synthesis of reduced graphene oxide zinc sulfide composite with enhanced photocatalytic activity. *Mater. Sci. Eng. B, 204*, 8–14, 2016.
32. Wojtyła, S. and Baran, T., Insight on doped ZnS and its activity towards photocatalytic removing of Cr(VI) from wastewater in the presence of organic pollutants. *Mater. Chem. Phys., 212*, 103–112, 2018.
33. Madkour, M. and Al Sagheer, F., Au/ZnS and Ag/ZnS nanoheterostructures as regenerated nanophotocatalysts for photocatalytic degradation of organic dyes. *Opt. Mater. Express, 7*, 1, 158–169, 2017.
34. Liu, S., Li, M., Li, S., Li, H., Yan, L., Synthesis and adsorption/photocatalysis performance of pyrite FeS2. *Appl. Surf. Sci., 268*, 213–217, 2013.

35. Mlowe, S., Osman, N.S., Moyo, T., Mwakikunga, B., Revaprasadu, N., Structural and gas sensing properties of greigite (Fe3S4) and pyrrhotite (Fe1-xS) nanoparticles. *Mater. Chem. Phys.*, *198*, 167–176, 2017.
36. Guo, X., Fu, Y., Hong, D., Yu, B., He, H., Wang, Q., Xing, L., Xue, X., High-efficiency sono-solar-induced degradation of organic dye by the piezophototronic/photocatalytic coupling effect of FeS/ZnO nanoarrays. *Nanotechnology*, *27*, 37, 375704, 2016.
37. Sakthivel, T., Venugopal, G., Durairaj, A., Vasanthkumar, S., Huang, X., Utilization of the internal electric field in semiconductor photocatalysis: A short review, *J. Ind. Eng. Chem.*, 72, 18–30, 2019.
38. Pouretedal, H.R. and Tavakkoli, M., Photodegradation of para-nitrophenol catalyzed by Fe2O3/FeS nanocomposite. *Desalin. Water Treat.*, *51*, 22-24, 4744–4749, 2013.
39. Rashid, J., Saleem, S., Awan, S.U., Iqbal, A., Kumar, R., Barakat, M.A., Arshad, M., Zaheer, M., Rafique, M., Awad, M., Stabilized fabrication of anatase-TiO2/FeS2 (pyrite) semiconductor composite nanocrystals for enhanced solar light-mediated photocatalytic degradation of methylene blue. *RSC Adv.*, *8*, 22, 11935–11945, 2018.
40. Lin, Y., Li, J., Chen, S., Zhou, H., Shu, Y., Tang, L., Long, Q., Zhang, P., Huang, Y., *In situ* construction of pyrite-marcasite-magnetite composite via FeS2 phase transformation and oxidation for the synergistic degradation of methyl orange and Cr (VI). *Sep. Purif. Technol.*, *308*, 122764, 2023.

4

Metal–Organic Framework as Potential Candidates for Photo-Electrocatalysis

Isha Soni[1] and Gururaj Kudur Jayaprakash[1,2*]

[1]*Laboratory of Quantum Chemistry, School of Advanced Chemical Sciences, Shoolini University, Bajhol, Solan, Himachal Pradesh, India*
[2]*Department of Chemistry, Nitte Meenakshi Institute of Technology, Bangalore, Karnataka, India*

Abstract
Photo-electrochemistry combines photocatalysis and electrocatalysis working synergistically. They improve the low selectivity of the electrocatalysts and solve the limitation of the photocatalyst, associated with poor charge carrier separation. A variety of metal oxides, metal sulfides, in combination with carbon materials, or a mixture of both have been studied for their application as a photo-electrocatalyst (photoelectrode). But MOFs, an emerging class of photo-electrocatalyst, subjugate metal oxides or sulfides in terms of their efficient mass and charge transport, high porosity, good absorption, and adsorption capacity, the vast number of active sites, and wide surface area. Moreover, the presence of metals in the framework makes them excellent photoactive semiconductors. In the current chapter, there is a comparative study of photocatalysis, electrocatalysis, and photo-electrocatalysis for sensing pollutants and CO_2 reduction.

Keywords: Catalysts, electrochemistry, redox, semiconductors

4.1 Introduction

Photocatalysis and electrocatalysis have revolutionized as the two important branches of catalysis in the scientific world. The science of photocatalysis is relatively new, but it has already proven to be an effective method

**Corresponding author*: rajguru97@gmail.com

Sadanand Pandey, Elvis Fosso-Kankeu and Soumya Pandit (eds.) Synergy of Bio-Chemical Processes for Photocatalytic and Photoelectrochemical Wastewater Treatment, (73–102) © 2024 Scrivener Publishing LLC

for purifying air, water, and surfaces. Photocatalysis has garnered a great deal of interest recently owing to its potential in providing clean and sustainable energy [1]. It is used for a wide range of tasks, including carbon dioxide remediation, air purification, pollutant degradation [2–4], and water splitting. By harnessing the power of sunlight, photocatalysts can efficiently convert solar energy into useful fuels and chemicals, providing a promising alternative to fossil fuels [5]. On the other hand, electrocatalysis is applied in several electrochemical processes, including fuel cells and batteries, and is essential for the advancement of renewable energy technology. It involves the use of a catalyst to facilitate electron transfer reactions, which can result in improved efficiency and performance in electrochemical devices [6, 7].

The development of new catalysts, and the optimization of existing ones are ongoing research fields in both the processes (photocatalysis and electrocatalysis). Researchers are continuously seeking to improve the efficiency, selectivity, and durability of catalysts, as well as to better understand the fundamental mechanisms involved in these processes. In terms of tackling the world's energy and environmental concerns, these two catalytic sectors are becoming increasingly significant. So, the current chapter brings forth light on the basic mechanisms of photocatalysis, electrocatalysis, and the photo-electrocatalysis. It also describes the various catalysts employed in these processes and how MOFs have proved an edge over other catalytic materials.

4.2 Photocatalysis

Photocatalysis is a combination of two chemical processes photochemistry and catalysis. Therefore, it is a light-driven catalytic reaction. According to the physical states of the photocatalyst and the reactants, photocatalysis may also be divided into homogeneous and heterogeneous types. When both the catalyst and the reactants are in the same phase, it is termed homogeneous photocatalysis; when both are in different phases, it is termed heterogeneous photocatalysis [8]. The main key components of photocatalysis include a photocatalyst, light source, pollutant, oxidizing agent, and supporting material. The photocatalyst, among others, is a crucial part of photocatalysis and will be covered in more detail in the later portion of the chapter. Photocatalytic reactions require light to activate the photocatalytic material and drive the reaction [9]. The type of light used depends on the photocatalytic material and the desired outcome. Common light sources include ultraviolet light, visible light, and infrared light. Whereas, the oxidizing agent is required to complete the photocatalytic reaction by accepting electrons from the photocatalytic material. This helps to maintain the charge balance in the reaction and promotes the degradation of the contaminant or pollutant. Supporting materials such as catalysts, co-catalysts, and sensitizers can be added to the photocatalytic system to enhance the performance of the photocatalytic reaction. For example, catalysts can speed up the reaction rate, whereas sensitizers can extend the absorption range of the photocatalytic material to allow for different light sources.

When a photocatalyst is irradiated with light, it absorbs light of energy higher than the bandgap energy, which leads to photo-excitation (generation of e^- and h^+ pairs by excitation of the electron from the valence band toward the conduction band). The reduction process is carried out by these conduction band electrons, whereas the oxidation process is carried out by the valence band holes for the surface-adsorbed species (Figure 4.1) [10].

$$hv + \text{photocatalyst} \rightarrow \text{photocatalyst}\ (e^- + h^+)$$

In simple words, a photocatalytic process can be thought of as a two-step reaction. First, the photocatalyst absorbs light and converts it into energy. This energy excites the electrons in the photocatalyst, which then causes chemical reactions to occur. During the second step, these reactions produce reactive species that are capable of breaking down organic pollutants or contaminants. Moreover, the efficiency of the photocatalyst is governed by the formation of e^- and h^+ pairs and the rapid recombination of these e^- and h^+ pairs [11, 12].

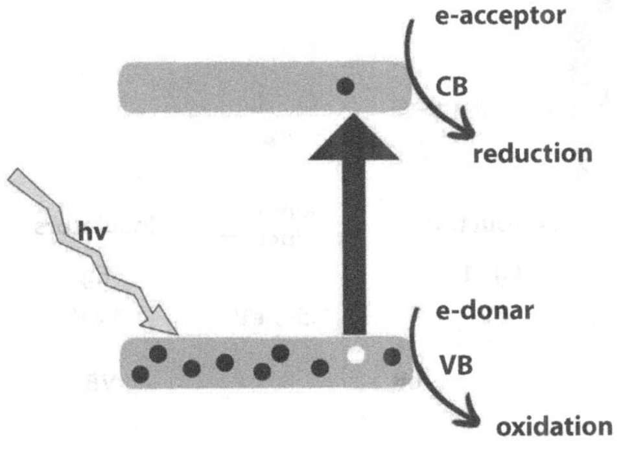

Figure 4.1 Redox mechanism occurring at the photocatalysis.

4.2.1 Photocatalyst

The term "photocatalyst" refers to any material that can absorb light and trigger chemical reactions. Examples of photocatalytic materials include titanium dioxide (TiO_2), zinc oxide (ZnO), and iron oxide (Fe_2O_3) [13]. The main function of the photocatalytic material is to absorb light and generate e^- and h^+ pairs, which drive the photocatalytic reaction [14].

Based on the bandgap between the valence band and conduction band (also known as highest occupied molecular orbital (HOMO) and lowest unoccupied molecular orbital (LUMO), respectively), materials are categorized into three types as shown in Figure 4.2: insulators, semiconductors, and conductors [15]. Out of these semiconductors are mostly employed in the photocatalytic activities.

Photocatalysts have been categorized into six classes: molecular photocatalysts concerning the molecular orbital theory, semiconductor photocatalysts concerning band theory, quantum dot photocatalysts when the quantum effect is observed, 2-D material-based photocatalysts concerning dimensionality, plasmonic metal-based photocatalyst, and photovoltaic-assisted catalysts [16, 17].

Advantages of photocatalysts:

- They are environmentally friendly, as they produce no harmful by-products and do not require the use of harmful chemicals.
- They are self-sustaining, as they can generate their energy from light. This makes photocatalysts ideal for use in remote

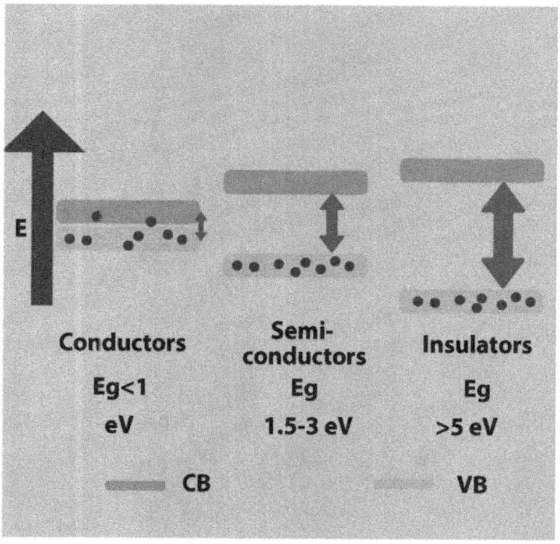

Figure 4.2 Different types of materials based upon bandgaps.

or low-energy environments where traditional methods of purification are not feasible.
- They are highly efficient.

Disadvantages of photocatalysis:

Photocatalysts are typically expensive to produce and are not widely available. This means that it may take some time for photocatalysts to become more widely used. Additionally, photocatalysts may not be effective against all types of pollutants, and further research is needed to develop more effective photocatalysts. Apart from these, there are a few more limitations of photocatalysis:

1. Narrow light absorption: Photocatalysts typically only absorb a limited range of the electromagnetic spectrum, which can limit their efficiency in certain applications. For example, some photocatalysts only absorb UV light, comprising only a smaller fraction of the solar radiation.
2. Catalyst degradation: The photocatalyst used in photocatalysis can become degraded over time, which can lead to reduced catalytic activity and decreased efficiency of the process.
3. Stability: Some photocatalysts can be unstable under certain reaction conditions, which can limit their lifetime and reduce their efficiency.
4. Environmental concerns: Some of the photocatalysts used in photocatalysis can be toxic or environmentally harmful, which can limit their use in certain applications.

Out of different classes of photocatalysts semiconductor photocatalysts are the most intensively explored. So, the current chapter mainly focuses on semiconductor photocatalysis. Here, the relative position of conduction and valence band minima is responsible for the redox potential of a catalyst and the position of fermi level with respect to conduction or valence band decides the type of a semiconductor formed as shown below (Figure 4.3).

To alter the bandgap of a semiconductor photocatalyst to improve its photocatalytic activity (in terms of increasing bandgap and increasing recombination period) [16, 18], two strategies have been used: i) homojunction and ii) heterojunction.

In semiconductor research, a homojunction is a junction between layers of similar semiconductors with equal bandgaps with different concentrations of dopants materials (p-p homojunction or n-n homojunction),

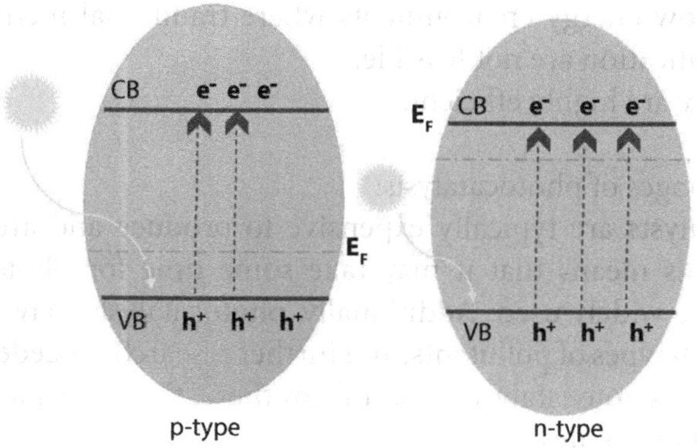

Figure 4.3 Types of semiconductors.

whereas heterojunction (Figure 4.4) is typically used to refer to a junction between two distinct monocrystalline semiconductor materials (p-n heterojunction or n-p heterojunction) [19, 20].

Heterojunctions are further classified into three main types (Figure 4.5): type I (straddling), type II (staggered), and type III (broken). In type I heterojunctions, compared to the corresponding bands in semiconductor B, semiconductor A's conduction band (CB) and valence band (VB) are, respectively, higher and lower. Under the influence of light, semiconductor B will accumulate e^- and h^+ at its CB and VB levels, respectively.

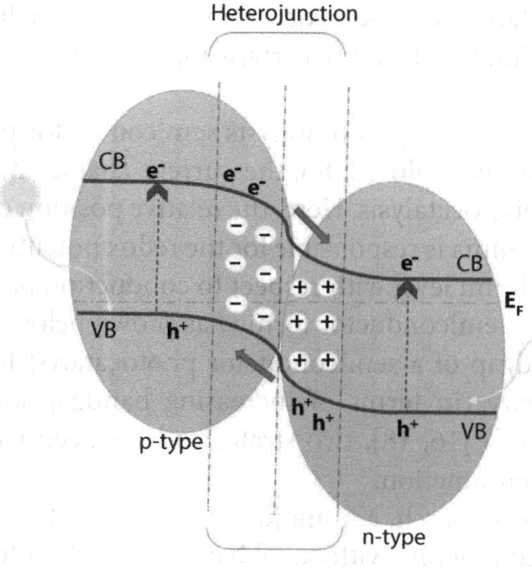

Figure 4.4 Heterojunction formed by p-n semiconductor showing space charge layer.

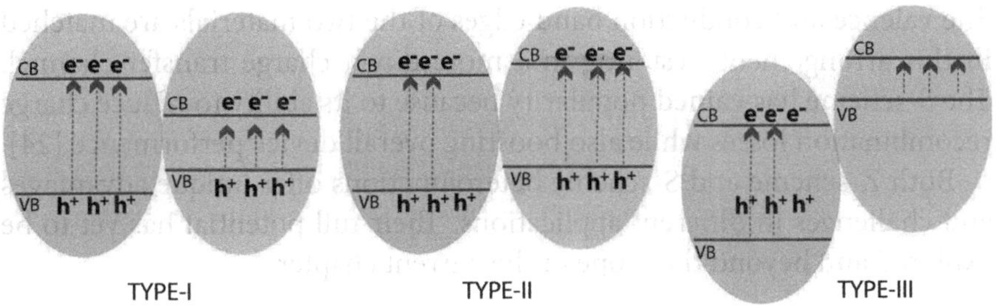

Figure 4.5 Different types of heterojunctions.

The e^- and h^+ pairs cannot be efficiently separated for the type I heterojunction photocatalyst because both electrons and holes collect on the same semiconductor. Moreover, a redox reaction occurs on the semiconductor with the lower redox potential, greatly diminishing the heterojunction photocatalyst's redox capability. Semiconductor A has higher CB and VB values than its counterpart values of semiconductor B for the type II heterojunction photocatalyst [21]. Consequently, during light irradiation, the photogenerated holes will migrate to semiconductor A while the photogenerated electrons will move to semiconductor B. This shifting causes an increase in the lifetime of the photoactive electron–hole pair. Similar to the type I heterojunction, the type II heterojunction photocatalyst's redox ability will be diminished since the reduction reaction and the oxidation reaction occur on semiconductors B and A, respectively, which have lower reduction potentials and lower oxidation potentials. The photocatalyst having type III heterojunction is comparable to the type II heterojunction, with the exception that the severe staggered gap prevents the bandgaps from overlapping. Hence, the type III heterojunction is not ideal for boosting the separation of e^- and h^+ pairs because e^- and h^+ migration and separation between the two semiconductors cannot occur for it [22].

Recent advancements in the field of semiconductor heterojunctions have led to the development of new types of Z-scheme and S-scheme heterojunctions, offering exciting prospects for various applications in the field of photonics and optoelectronics.

Two distinct semiconductor materials are coupled to form a heterojunction, and their energy bands are staggerly aligned in Z-scheme heterojunction. This configuration allows for effective charge separation and transmission over the interface by generating a cascade of energy levels [23]. The S-scheme heterojunction, on the other hand, illustrates an alternate energy level alignment in which the energy bands of the two materials are directly aligned without the necessity for a staggered structure.

The valence and conduction band edges of the two materials are matched in this arrangement, resulting in a more direct charge transfer channel. The S-scheme has gained popularity because to its ability to reduce charge recombination losses while also boosting overall device performance [24].

Both Z-scheme and S-scheme heterojunctions offer unique advantages and challenges in different applications. Their full potential has yet to be explored and beyond the scope of the current chapter.

4.3 Electrocatalysis

It is a heterogeneous catalytic process of converting electrical energy into chemical forms at the electrode interface [25]. Typical electrochemical setup is shown in Figure 4.6. Here, electrons flow from the electrode to the substrate, facilitating the redox process. Overall, in the electrocatalytic processes, the presence of a catalyst (electrocatalyst) on the electrode surface enhances the reaction rate and selectivity, allowing for more efficient energy conversion and better utilization of resources [26].

Various components of electrocatalysis are as follows:

Electrode: This is a conductive material that serves as the site of electron transfer between the chemical reaction and the external electrical potential.
Electrolyte: This is a medium that conducts ions and allows for the transfer of charge between the electrodes.

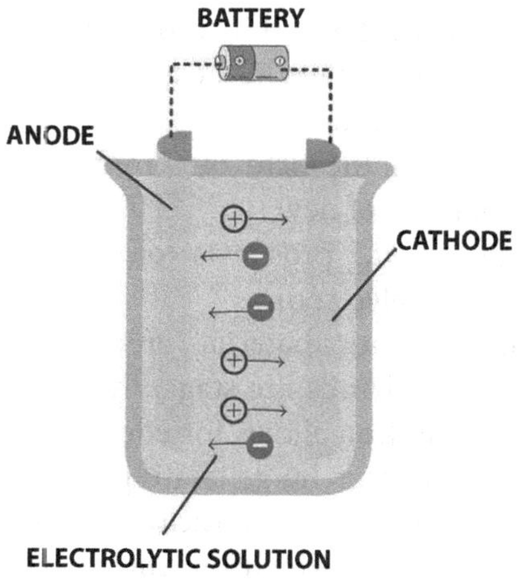

Figure 4.6 Typical electrochemical cell.

Catalyst: The term "catalyst" refers to a material that speeds up a chemical reaction without being consumed in the process. In electrocatalysis, the catalyst is typically located at the electrode surface and facilitates the transfer of electrons between the electrode and the reactants.
Power source: This is an external source of electrical potential that drives the reaction by providing the energy needed to overcome activation barriers.
Reactants: These are the starting materials that transform the presence of the catalyst and electrical potential.

The typical mechanism for the electrochemical process is shown in Figure 4.7 and consists of mainly four steps:

Adsorption: The substrate molecule adsorbs on the surface of the catalytic electrode, forming an intermediate species that is stabilized by the catalyst.
Electrochemical activation: The substrate molecule undergoes an electrochemical reaction with the electrode, which results in the flow of electrons between the substrate and the electrode.

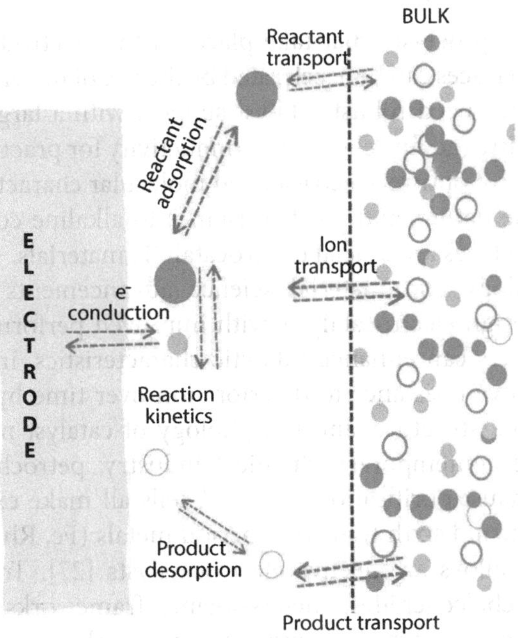

Figure 4.7 General electrochemical mechanism.

Charge transfer: The catalytic site at the electrode surface serves as an intermediate between the substrate and the electrode, assisting in the ease of electron flow between the substrate and the electrode. Surface reaction: The adsorbed substrate species on the catalytic site undergoes a surface reaction, resulting in the formation of the desired product.
Product desorption: The product is desorbed from the surface of the electrode and dissolves in the electrolyte solution.

The function of each component is critical to the overall operation of an electrocatalytic system.

In short, the key to efficient electrocatalysis is the ability of the catalyst to adsorb and activate the substrate, facilitate charge transfer, and promote the surface reaction leading to product formation.

The electrode acts as the interface between the chemical reaction and the electrical potential, whereas the electrolyte provides a medium for ion transport. The catalyst participates in the reaction by lowering the activation energy required to trigger the reaction, whereas the power source provides the energy needed to drive the reaction. Finally, the reactants and products are the materials involved in chemical transformation.

4.3.1 Electrocatalyst

Electrochemical processes that take place on the electrode surface or at solid/liquid interfaces can be accelerated by the use of materials called electrocatalysts. They need robust catalyst support with a large surface area, chemical stability, and fine electronic conductivity for practical usage. Due to their distinctive optical, electrical, and molecular characteristics, highly functionalizable nature, and great resistance to alkaline corrosion, metal oxide nanostructures are crucial electrocatalytic materials.

Nanotechnology and material science advancements have enabled scientists to design electrocatalysts with improved performance and stability. Researchers can enhance catalytic characteristics, increase surface area, and improve resistance to deterioration over time by manipulating the composition, structure, and morphology of catalyst materials at the nanoscale. The contemporary chemical industry, petrochemical industry, vehicle exhaust purification, and fuel cells all make extensive use of the fourth, fifth, and sixth transition period metals (Fe, Rh, Pd, Ir, Pt, Au, etc.) and their alloys as effective electrocatalysts [27]. Transition metal oxides, metal chalcogenides, metal–organic frameworks (MOFs), and carbon-based materials such as graphene and carbon nanotubes are all being investigated by researchers.The pace of electrochemical processes can be greatly accelerated using electrocatalysts. At decreasing potential levels, the rate of electron flow between the electrode surface and the reactants of the electrochemical process rises. Nearly invariably, a more effective electrocatalyst will have a lower potential for electrochemical processes. Electrochemical reactions include overall water splitting, HER at the cathode, and oxygen evolution.

4.4 Photo-Electrocatalysis

Photo-electrocatalysis is a hybrid process that combines the principles of photocatalysis and electrocatalysis to generate chemical reactions [28]. Therefore, photo-electrocatalysis refers to the use of light-activated catalysts to drive a redox reaction. In photo-electrocatalysis, a semiconductor material acts as a photocatalyst and an electrode simultaneously [29].

The various components of photo-electrocatalysis are as follows:

1. Photoanode: A substance that produces electron–hole pairs after absorbing light energy is a photoanode. The electrons can participate in the redox reaction, whereas the holes can oxidize the target species.
2. Electrolyte: This an electrically conductive solution that provides a medium for the movement of ions between the photoanode and the cathode. It also acts as a source of reactants and products.
3. Cathode: This is a material that receives electrons from the photoanode and reduces the target species. It can also serve as an electron sink to prevent back reaction.
4. Light source: This provides the energy to activate the photoanode and drive the redox reaction. The light source can be natural or artificial, and it should have the appropriate wavelength to match the bandgap of the photoanode
5. Catalyst: This is a material that can increase the reaction rate and selectivity by lowering the activation energy. It can be deposited on the photoanode or cathode to enhance the reaction kinetics.

These components work together as shown in Figure 4.8 to drive a redox reaction under light illumination, where the photoanode generates electrons and the cathode receives electrons to reduce the target species [30].

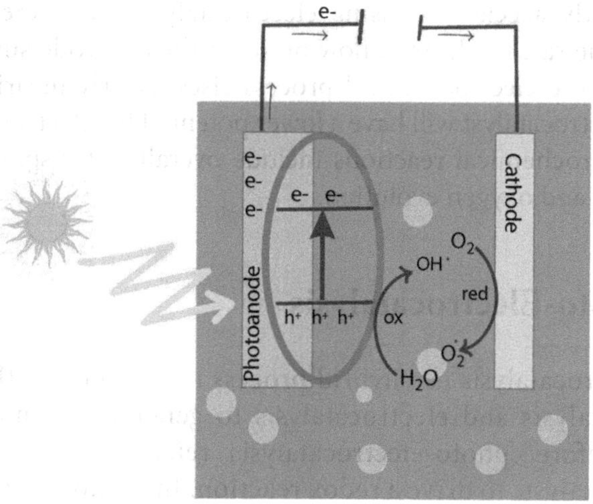

Figure 4.8 General photo-electrocatalytic mechanism.

Overall, the three main processes occur during the photo-electrocatalysis as follows [31]:

a. Light absorption: Light is captured by the semiconductor material, which then generates electron–hole pairs. Positive holes are left in the valence as the electrons are stimulated to the conduction band.
b. Electron transfer: To create a current flow, the electrons in the conduction band are moved to the electrode, whereas the holes in the valence band take part in the chemical process.
c. Chemical reaction: Adsorbed substances on the surface interact with the holes in the semiconductor's valence band to create reactive species like hydroxyl radicals or superoxide ions. These species can then react with other species in the solution to generate the desired products.

Therefore, the overall efficiency of photo-electrocatalysis depends on several factors, including the light-absorption properties of the semiconductor, the potential difference between the electrode and the solution, and the surface area of the electrode. The reaction can be optimized by controlling these parameters and by choosing the appropriate semiconductor material and co-catalyst [32].

4.4.1 Photo-Electrocatalyst

In the case of photo-electrocatalysis, the photoanode acts as a photo-electrocatalyst. So, to perform photo-electrocatalysis, further study has to be done on photoanode preparation materials used as photoelectrodes. Therefore, the right semiconductor material used in the photo-electrocatalysis process needs to satisfy several criteria, including a broad light absorption range of the spectrum, adequate band energies, strong charge transfer, excellent catalytic activity, good flexibility, sustainability, and affordability. Various other factors that affect the photo-electrocatalytic performance include the stability and film thickness of photoelectrodes [33, 34]. Some of the most commonly employed materials as photo-electrocatalyst are described below.

1. TiO_2: TiO_2 is one of the most widely studied photo-electrocatalysts. TiO_2 can only absorb UV radiation with a wavelength below 385 nm because of its bandgap, which is around 3.2 eV. It promotes the reduction of water to hydrogen gas. With its strong UV oxidation capabilities, low price, great stability, and lack of toxicity, TiO_2 is employed in the photoelectrochemical (PEC) degradation of organics in aqueous settings. To enhance PEC performance, many TiO_2 films, including nanotube arrays and nanorod films, have been developed. TiO_2 nanotubes with varying tube lengths have been produced using anodization, which has shown enhanced light absorption due to increased surface area. Several strategies have been used to increase the visible light absorption of TiO_2 and reduce the recombination of the photogenerated charge carriers.
2. ZnO: ZnO is another semiconducting material that can be used as a photo-electrocatalyst. With strong photocatalytic activity equivalent to TiO_2, ZnO is a semiconductor with similar properties. It works well to encourage the breakdown of organic contaminants in water. It has a band gap of 3.0 eV, which is a bit less than TiO_2's bandgap, characterizes it. Nanowires [35], nanorods [36], and nanoflakes [37] are the three types of ZnO films that have been most often reported.
3. Just 5% of the solar radiation spectrum is UV light. A focus of the intense study is now being given to the creation of visible light-responsive film electrodes to fully use visible light, which accounts for 43% of solar light. Binary and ternary

metal oxides such as WO_3, $BiVO_4$, and Bi_2WO_6, along with organic semiconductors like g-C_3N_4, are some of the most researched visible light-responsive films. Having a bandgap of 2.4 eV, $BiVO_4$ has been employed in environmental cleanup and water splitting by removing a wide range of organics. Metal-doped $BiVO_4$ has been shown to have enhanced PEC properties compared to undoped $BiVO_4$. Graphene and other graphene-based materials have been studied for use as photo-electrocatalysts due to their unique optical and electronic properties. They have the potential to be employed in a variety of applications, like pollution degradation, CO_2 reduction, and water splitting.

4. Another emerging photo-electrocatalyst is MOF. It has been further elaborated in detail in a further section of the chapter.

4.4.2 MOFs

MOFs are a subclass of porous crystalline materials made up of organic linkers and metallic nodes (such as metal ions or clusters) [38]. MOFs have lately drawn more interest in the realm of photocatalysis due to their extremely high specific surface area, diverse topology, and readily modifiable porous structure. MOFs are molecules with crystalline lattice structures with delocalized conduction bands and valence bands, which distinguishes them from traditional inorganic semiconductors.

They also possess semiconductor behavior where the metal-oxo clusters behave as quantum dots while the organic linkers/ligands act as light-absorbing antennas [39]. In other words, they can be considered isolated semiconductor quantum dots [40]. The main processes responsible for their photocatalytic properties are ligand-to-charge transfer (LMCT), metal-to-ligand charge transfer (MLCT), and ligand-to-ligand charge transfer (LLCT). Out of these, LMCT is the most common.

The prerequisite condition for the photocatalysis is light responsiveness of the MOF [41, 42].

For the excitation of MOF:

$$E_{hv} = 1240/\lambda > E_{HOMO} - E_{LUMO}$$

$$\text{UV light}: \lambda < 365 \text{ nm or } E_{hv} > 3.1 \text{ eV}$$

$$\text{Visible light}: \lambda > 400 \text{ nm or } E_{hv} < 3.1 \text{ eV}$$

Advantages of MOF as a photocatalyst over the traditional metal oxide semiconductors:

1. Large surface area
2. Porous structure
3. Reduced metal-to-metal interaction
4. MOFs offer a significant chance to expand the light response across a wide range due to their structural tunability.
5. Its volume recombination is significantly reduced because of its crystalline form, which has minimized structural flaws that are generally hotspots of e^- and h^+ recombination.

Here in the case of the MOF, the HOMO-LUMO bandgap is controlled by the sp^2-hybridized state of C of the organic linker molecule. Thus, the organic linker harvests the light radiations falling into the crystal lattice of the MOF, leading to the photoexcitation as shown in Figure 4.9 (generation of the electron–hole pairs). This photoactivation is passed to the metal nodes by various charge transfer processes as discussed before, thereby activating the metal nodes by bringing changes in the electronic states of the metal [43]. Metal–organic frameworks (MOFs) have demonstrated significant promise in photo-electrocatalysis, but a number of problems may restrict their involvement in this sector. These include inherent bandgaps, inefficient charge transport, restricted active sites, mass transport limits, and scalability and cost constraints. Various approaches that have been employed to increase the photocatalytic efficiency of the MOFs including ligand engineering, photosensitizer inclusion, and co-catalyst deposition (shown in Figure 4.10). Ligand engineering is the process of altering ligands

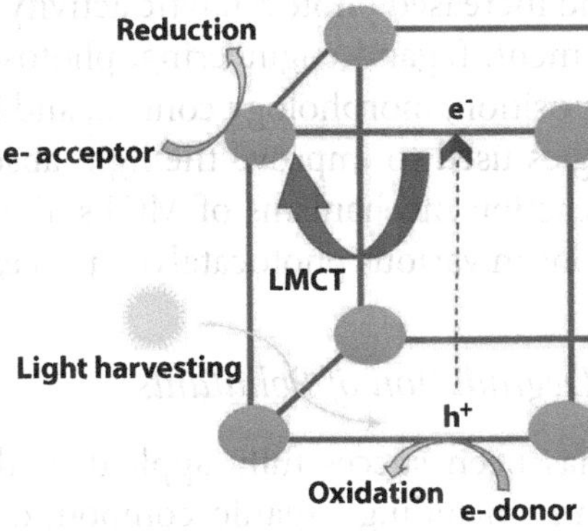

Figure 4.9 Photocatalytic mechanism occurring at the MOF crystal.

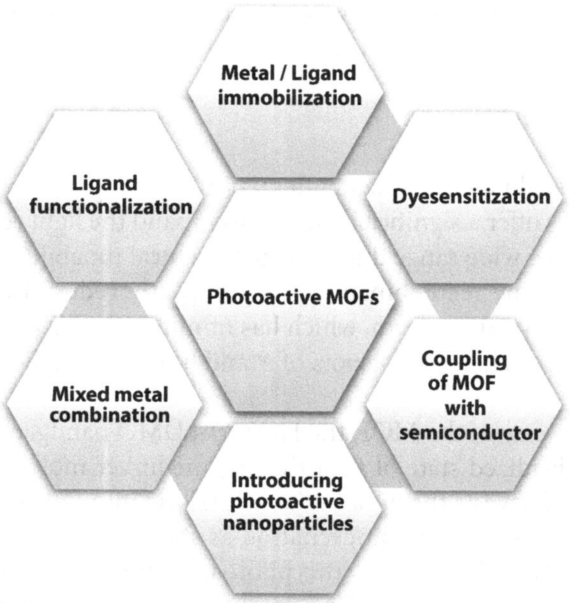

Figure 4.10 Various strategies to improve MOFs photocatalysis efficiency.

in order to tailor the energy levels and electrical characteristics of a MOF, whereas photosensitizer inclusion is the process of introducing photosensitizers into the MOF structure or depositing them on the MOF surface. Co-catalyst deposition is the process of facilitating charge transfer and catalytic reactions at the MOF surface. Deposition of co-catalysts onto a MOF, such as metal nanoparticles or metal oxides, can dramatically improve its photocatalytic activity. Morphology adjustment can change the size, shape, and surface area of the MOF, resulting in increased photocatalytic activity. Synergistic effects and increased photocatalytic activity can result from heterostructure development. Ligand engineering, photosensitizer incorporation, co-catalyst deposition, morphology control, and heterostructure formation are all strategies used to improve the light absorption, charge transfer, and catalytic reaction mechanisms of MOFs, thereby increasing their potential applications in various photocatalytic processes [39, 44].

4.4.2.1 *Sensing and Degradation of Pollutants*

Photo-electrocatalysis has been successfully applied in the degradation of a variety of pollutants, including organic compounds such as dyes, pharmaceuticals, and pesticides, as well as heavy metals, such as lead and cadmium. It offers several advantages over other degradation methods, including high efficiency, low cost, and the ability to degrade a wide range of pollutants. This process can also be carried out under ambient conditions, without the need for additional heat or pressure.

In this process, a photocatalyst is activated by light, which generates electron–hole pairs that can then be utilized for pollutant degradation [45]. The generated electrons move toward the electrode, whereas the holes migrate to the surface of the photocatalyst. The electrons are then utilized for the reduction of the pollutant or the generation of reactive oxygen species (ROS) that can also degrade pollutants. On the other hand, the holes react with water to produce hydroxyl radicals (OH•), which are powerful oxidants that can degrade pollutants [46, 47].

$$\text{Semiconductor} + h\nu\ (h\nu > E_g) \rightarrow \text{Semiconductor}\ (e_{CB}^- + h_{VB}^+)$$

$$h_{VB}^+ + H_2O \rightarrow \bullet OH + H^+ + e^-$$

$$h_{VB}^+ + OH^- \rightarrow \bullet OH$$

$$e_{CB}^- + O_2 \rightarrow O_2\bullet^-$$

$$e_{CB}^- + h_{VB}^+ \rightarrow \text{Semiconductor} + \Delta$$

To maintain environmental security, it is crucial to monitor the levels of the organo-arsenic pollutant p-arsanilic acid. Using a porphyrin-based MOF on r-graphene oxide (PCN-224/rGO) as electrode material with photocatalytic activity as well as binding capabilities has been evaluated as a photo-electrochemical sensor for p-arsanilic acid detection [48]. The mechanism analysis showed that the PCN/rGO exhibited a great affinity for p-ASA using the Zr-O-As bond as well as π–π stacking of graphene. Further, on adsorption of p-ASA, the photoelectrode showed only a slight change in the bandgap (0.02 eV), resulting in a shorter depletion layer and a lower electric field intensity, but photocurrent increased. The [tetrakis(4-carboxyphenyl)porphyrin] (TCPP) in PCN-224 is thought of as a light-harvesting component where electrons are pushed by photoexcitation from the valence band to the conduction band. Furthermore, photogenerated electrons may convert the dissolved oxygen to a superoxide radical $O_2^{\bullet -}$, enhancing the cathodic photocurrent. Therefore, in addition to having strong binding properties, this nanocomposite displayed high photoelectric activity with a lower detection limit of 5.47 ng L^{-1}.

A photo-electrochemical sensor made of Cu-BTC/g-C3N4 nanosheet was developed. When utilized as a composite material, the Cu-BTC traps the insecticide molecules and speeds up the signal transfer, whereas CN-NS has great optical performance. When exposed to visible light radiations, the metal (Cu) center mixes with the additional glyphosate to form metal-glyphosate complexes, which raise the steric barrier for electron flow and reduce photocurrent. The designed sensor can detect glyphosate using this technique in both electroactive and non-electroactive modes. The results demonstrate that this Cu-BTC/g-C3N4 nanosheet-based sensor has a large detection range (1×10^{-12} to 1×10^{-8} and 1×10^{-8} to 1×10^{-3}) with 0.13-pM detection limit. The photocurrent is significantly reduced after adding nanoparticles to the electrode surface. By creating a connection between the relative change in photocurrent and the concentration of glyphosate, this sensor enables sensitive PEC detection of non-electroactive glyphosate. According to the experimental findings, this porous Cu-BTC MOF/CN-NS–based photo-electrochemical sensor has a broad detection range and a low detection limit for glyphosate, which is not an electrically active substance. This work shows that porous MOF material may function as a novel hybrid material for photo-electrochemical detection [49].

By covering powdered catalysts over the conductive glass, conventional photoanodes reduced the rate of mass transfer. A photo-electrocatalyst comprising of semiconductor and MOF hybrid was effectively created using the *in situ* growing approach [50]. As compared to TiO_2-NTs without modifications, ZIF-8/NF-TiO_2 heterogeneous photo-electrocatalyst demonstrated several advantageous traits such as increased photocurrent responsiveness, enhanced charge ejection efficiency, and decreased charge transfer resistance are all results of charge creation and separation, such as the construction of an anatase rutile heterojunction. This led to increases in ZIF-8/NF-reaction TiO_2's rate and degradation efficiency of 21.7 times and 11.6 times, respectively. The improved photoanode seems to have outstanding stability based on the results of cyclic runs. Cleavage, the opening of the aromatic ring, and mineralizing are the three primary processes for the degradation of sulfamethazine. The following explanations were given for ZIF-8/NF-superior TiO_2's PEC activity: i) Greater light utilization may be caused by N and F doping; ii) the double heterojunction structure between ZIF-8, rutile phase, and anatase effectively promoted carriers transfer; and iii) the doping of N and F could broaden the absorption edge to the visible light region and promote the formation of N-Ti-O bonds, which suppressed the recombination of e^-/h+ pairs.

Even MOF nanorods have been employed as a sacrificial template for synthesizing photoelectrode material. Also, this provides a good understanding of the synthesis of photocatalysts based on unique MOFs. Fe-MOF nanorods were MOF nanorods as a template as well as a precursor during the thermal carbonization process under N_2 atmospheres to create ternary TiO_2@C/$FeTiO_3$ (TCF) hollow nanotubes [51]. The hollow TCF composites can efficiently harvest full spectrum light energy, improve interfacial charge separation, and suppress the recombination of photogenerated e^- and h^+ pairs due to their delicate structural features, which include a hollow needle-like framework with conductive- carbon layer having $FeTiO_3$ nanoparticles. This results in enhanced photo-electrocatalytic activity. By varying Fe/Ti molar ratio in the reaction solution, it is possible to govern the structural characteristics that are associated with the catalytic capabilities of TCF hollow nanotubes, with the shell thickness of these tubes being controllably regulated between 20 nm and 25 nm. Among the synthesized samples, the TCF-20% photoanode performs the best at degrading phenol when irradiated with light along with an anodic bias of 1.5 V vs. SCE. It gives a degradation rate of 0.586 h^{-1}, which is 2.75 times more than that of the equivalent of photocatalysis and electrocatalytic processes.

4.4.2.2 CO_2 Reduction

The mitigation of the global warming problem and closing of the enormous gap between energy demand and sustainable supply may be accomplished simultaneously through the conversion of CO_2 into value-added goods, which is a promising technique [52, 53]. Several products are produced when various photocatalysts are utilized in CO_2 conversion, including CH_4, CH_3OH, CO, HCOOH, and HCHO as shown in Figure 4.11 [54].

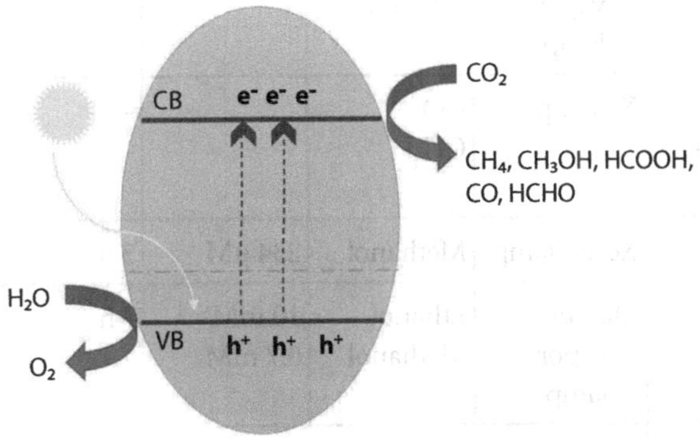

Figure 4.11 Showing various processes occurring at the CB and VB during CO_2 reduction at the photocatalyst surface.

$$CO_2 + 8H^+ + 8e^- \rightarrow CH_4 + 2H_2O \; E^0 = -0.240 \text{ V}$$

$$CO_2 + 6H^+ + 6e^- \rightarrow CH_3OH + H_2O \; E^0 = -0.380 \text{ V}$$

$$CO_2 + 2\,H^+ + 2e^- \rightarrow HCOOH \; E^0 = -0.610 \text{ V}$$

$$CO_2 + 2\,H^+ + 2e^- \rightarrow CO + H_2O \; E^0 = -0.530 \text{ V}$$

$$CO_2 + 4H^+ + 4e^- \rightarrow HCHO + H_2O \rightarrow E^0 = -0.480\text{V}$$

Photo-electrocatalysis may achieve the quick transfer of electrons to decrease CO_2 on the photocathode at a lower overpotential than the photocatalytic CO_2 reduction process [55], significantly improving reduction efficiency. Photo-electrocatalysis, as opposed to single electrocatalysis, uses sunlight as the energy source to stimulate the production of charge carriers and then takes part in the redox reaction to eventually achieve CO_2 reduction with lesser energy utilization [56]. Various MOFs have been investigated photo-catalytically [57], but only a few MOFs have been photo-electrochemically tested for CO_2 reduction (Table 4.1). The CO_2 reduction is a three-step process: i) Photoactivation and adsorption of CO_2 at the catalytic surface, ii) C-C bond formation, and iii) influence of

Table 4.1 Tabulation of MOFs used photoelectrochemically for CO_2 reduction.

Complex	Irradiation used	Conversion product	Conversion rate	Conversion time	References
ZIF-8, deposited on TiO2 nanotubes	Mercury vapor lamp	Methanol Ethanol	-	3 h	[59]
Ni2+ doped NH2-MIL-125-Ti	Xe lamp	CO CH_4	-	-	[60]
Cu(BDC)	Xe arc lamp	Methanol	234 μM	3 h	[61]
Ti/TiO_2NT-ZIF-8	Mercury vapor lamp	Ethanol Methanol	10 mM 0.7 mM	3 h	[62]
Ti/TiO_2NT-$Ru_3(BTC)_2$	Xe arc lamp	Methanol	314 μM	3 h	[63]

reaction conditions. The primary influencing factors in the photo-electrocatalytic reduction of carbon dioxide, such as the photoelectrode's influence (thermodynamics and kinetics), the interaction between catalysts and reaction molecules (initial activation, along with C-C bond formation), and reaction conditions (such as electrolyte effect, overpotential, and light source) [58].

The role of ascorbic acid in the photo-electrocatalytic conversion of CO_2 to alcohol is investigated [59]. The results prove that adding an electron donor agent to the electrolyte boosted CO_2 conversion to methanol by 22% and ethanol conversion by 12%, after 3 h of photo-electrolysis using UV-Vis irradiation on TiO_2 nanotubes coated with ZIF-8 nanoparticles under Eapp = 0.1V. This result is due to a more stringent restriction on all photogenerated charge carrier recombination, which is essential for promoting the reduction processes of carbamate compounds formed on the electrode interface. The study's findings offer a fascinating perspective on how immensely useful things can be created while also causing environmental damage.

In MOFs, controlling the electronic structure of metal oxo clusters is a viable strategy for adjusting charge transfer effectiveness and photocatalytic performance. An *in situ* doping technique has been used to synthesize a series of Ni^{2+} doped MOFs with the formula NH_2-MIL-125-Ni_x/Ti varying in Ni^{2+}/Ti^{4+} molar ratios from 0.5% to 1.5% [60]. Based on experimental and computational methods, correlations between the electronic structure of the Ti/Ni)$_8$O$_8$(OH)$_4$ nodes and bandgap, charge transfer effectiveness, and energy location of band edges of the NH2-MIL-125-Nix/Ti are comprehensively examined. It was established that the doped Ni^{2+} was a reliable mediator for facilitating the transport of electrons from the photoexcited ligand (terephthalate) to the metal oxo nodes [(Ti/Ni)$_8$O$_8$(OH)$_4$] in NH_2-MIL-125-Ni_x/Ti. As discussed, earlier electron flow from ligand to metal is a crucial step in photocatalysis, so the introduction of Ni^{2+} causes HOMO and LUMO to form more immediate states, which makes it easier for electrons to move from LUMO to HOMO but, as a consequence, may hasten the recombination process even. But results showed better performance in terms of 98.6% of selectivity for CO, which indicates that greater charge carriers are created with increased recombination span.

Even electrochemical deposition of MOF on electrode as thin films have been used for CO_2 reduction such as Cu-BDC on metallic copper to create the Cu/Cu2 O/Cu-BDC electrode, using 1,4-benzene dicarboxylate (1,4-BDC) as the ligand [61]. The size of the produced microcrystals and the thickness of the Cu-BDC thin films are both significantly influenced by the current density and electrodeposition time. Upon UV-vis irradiation, the photo-reduction of CO_2 for 3 h, an applied voltage of + 0.10 V, and 0.1 M aqueous sodium sulfate solution as supporting electrolyte resulted in a concentration of 234 μM methanol. Cu/Cu_2O-Cu(BDC) electrodes fabricated at 110°C, with electrodeposition times of 6.5 min and 30 min, produced the greatest photocatalytic activity. Studies such as diffuse reflectance infrared Fourier transform spectroscopy (DRIFTS) and Fourier transform infrared spectroscopy (FTIR) showed that the hybrid electrode's MOF component successfully captured CO_2. The results showed that using MOF enhanced CO_2/methanol conversion by around 20 times compared to using an electrode made of Cu/Cu_2O without the ligand. This happened as a consequence of the enhanced surface area provided by MOF for the better adsorption of CO_2.

Growing nanoparticles on Ti/TiO_2NT led to the creation of MOF-based hybrid catalysts, which were then utilized as a photo-electrocatalyst for CO_2 reduction reactions in an aqueous solution [62]. Analysis using FTIR spectroscopy revealed that both the pore structure and the chemical makeup of the MOF linker affected host–guest interactions. The composite, which was made by depositing ZIF-8 nanoparticles on TiO_2 nanotubes, effectively converted CO_2 into methanol and ethanol fuels by photo-electrocatalytic means. As a result, the zinc-based MOF served as a co-catalyst for photo-electrocatalytic reduction as well as a catalyst for CO_2 adsorption/activation. The addition of ZIF-8 to the Ti/TiO_2 electrodes boosted the photocurrent, and the values were significantly increased in the CO_2-saturated solution. Spectroscopic and voltammetric experiments showed that the CO_2 adsorbed at the photo-electrode surface generated stable photoactive carbamates. During UV-vis irradiation at ambient temperature, photo-electrocatalytic reduction of CO_2 at Ti/TiO_2NT-ZIF-8 electrodes produced up to 10 mM of ethanol and 0.7 mM of methanol. The photo-electrocatalysis out-shown the photocatalysis in terms of conversion efficiency.

$Ru_3(BTC)_2$ MOF and Ti/TiO_2 nanotubes were used to build a composite photo-electrocatalytic device [63]. On anatase TiO_2NT, a thin MOF coating was grown using a solvothermal technique, resulting in a uniform distribution and no obstruction of the nanotube surface. The hybrid material's inclusion of the ruthenium MOF significantly enhanced CO_2 adsorption on the electrode surface in addition to increasing visible light absorption. The CO_2 adsorbed on the MOF structure receives a direct transfer of the photogenerated electrons, which are trapped by the $Ru_3(BTC)_2$, resulting in the formation of $CO_2^{\bullet-}$. The negative potential provided can inhibit the VB holes or make them available for water oxidation, which produces hydrogen and oxygen ions that are essential for fostering the synthesis of alcohol. Due to the $Ru_3(BTC)_2$ MOF's dual role as a co-catalyst and its ability to transmit excited electrons for the reduction of CO_2 to methanol without recombination of e^- and h^+ couples, this resulted in quick electron transfer. This was amply demonstrated by comparing the electrode performances in the photocatalytic and photo-electrocatalytic procedures. The $TiO_2NT\text{-}Ru_3(BTC)_2$ electrode was used in the photo-electrocatalytic procedures, which produced the greatest results at −0.5 V vs. Ag/AgCl and generated 314 μM of methanol after 3 h of testing. The results supported the TiO_2NT electrode's strong performance and stability following modification with the Ru-based MOF, which considerably improved the photo-electrocatalytic reduction of CO_2 to methanol.

4.5 Comparison of Photocatalysis, Electrocatalysis, and Photo-Electrocatalysis

The comparison of photocatalysis, electrocatalysis and photocatalysis on the basis of cost, energy consumption, selectivity and various other parameters has been done below in the Table 4.2 [65–69].

Table 4.2 Advantages and disadvantages of photocatalysis, electrocatalysis and photocatalysis.

	Photocatalysis	Electrocatalysis	Photo-electrocatalysis
Advantages	• Less energy consumption • Higher catalyst stability • Low cost • Shorter electron transfer path • Ecofriendly	• Higher sensitivity • Higher selectivity	• Cost-effective • Higher sensitivity • Higher selectivity
		• Products formed after the reaction can be transformed into various energy sources • Easy scale-up • Higher electroactive surface area	• Products formed after the reaction can be transformed into various energy sources • Increases recombination period • Low background current • Easy operation

(*Continued*)

Table 4.2 Advantages and disadvantages of photocatalysis, electrocatalysis and photocatalysis. (*Continued*)

	Photocatalysis	**Electrocatalysis**	**Photo-electrocatalysis**
Disadvantages	• Separation of the catalyst after the reaction is a very tedious process • Complex electron transfer mechanism • Lowered selectivity • Efficiency highly dependent on light intensity • Scale-up is a costly affair	• Longer electron transfer pathway • Short-lived catalyst • Costly • The use of various acids and alkalis is required	• Complex construction • Efficiency highly dependent on light intensity • Greater chances of photo-corrosion

4.6 Conclusion and Future Aspects

Photoelectrochemical (PEC) detection is a prospective use in environmental monitoring because, in comparison to current methods, it is less expensive and simpler to use. The electrons in this process are photoexcited by photoexcitation, and, then, they move toward the electrode surface with the aid of externally applied voltage. To encourage the directed transfer of photogenerated e^- and h^+ and improve redox capacity, the action of an external electric field is advantageous. When the photocatalyst's band position is unfavorable for CO_2 reduction and H_2O oxidation, the redox potentials can be adjusted by applying the correct bias voltage to the photoreduction system. With a decreased overpotential, photo-electrocatalysis may quickly transfer electrons to reduce CO_2 on the photocathode, significantly improving reduction efficiency. However, some challenges remain in the commercialization of PEC, including optimizing the process for large-scale applications, reducing the cost of catalysts, and improving the stability and durability of the materials used in the process.

CO_2 reduction using PEC cell is very advantageous in terms of generating value-added products that are commercially very important due to their cost effectiveness and eco-friendly approach.

Overall, PEC offers great promise as an effective and sustainable method for pollutant degradation, with potential applications in the remediation of contaminated water and air, as well as in the production of clean energy.

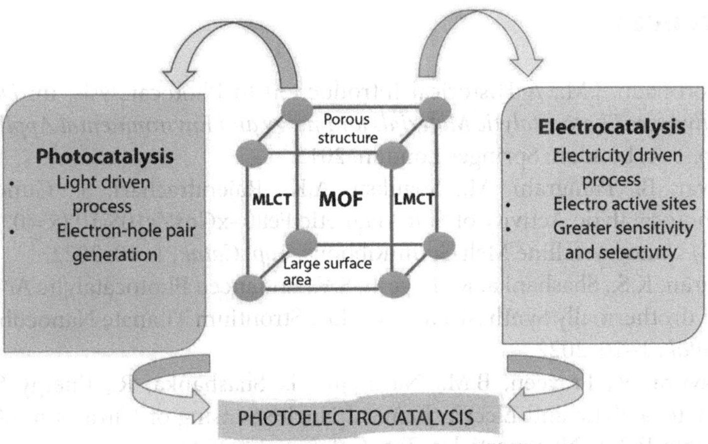

Figure 4.12 Diagrammatic representation of MOF properties for photo-electrocatalysis.

Several studies have demonstrated the potential of MOFs (Figure 4.12) in PEC for the degradation of various pollutants, such as organic dyes, phenol, and heavy metals. For example, one study reported the use of a MOF-based PEC system for the degradation of methylene blue, a common dye used in textile industries, and achieved high degradation rate. One advantage of MOFs in PEC is their ability to enhance the separation of charge carriers, which can increase the efficiency of the degradation process. MOFs can also be modified with various functional groups, such as metal or organic catalysts, to further enhance their catalytic properties.

Another advantage of MOFs in PEC is their potential for scalable production and easy separation from the treated water or air. However, further research is needed to optimize the synthesis and functionalization of MOFs for PEC applications, as well as to address the stability and durability issues that may arise under harsh operating conditions.

PEC sensors' usual electrode materials must be changed with biological probes to prevent problematic sensor design and recognizer deactivation after storage, as they lack unique analyte binding sites.So the employment of the bio-MOFs can be advantageous as MOFs are very potential candidates having large surface areas and porous structures with tunable structure and topology.

Future research will focus primarily on improving film electrode PEC performance, increasing MOF loading, and expanding MOF applications. The newly reported approach could facilitate the development of novel MOF-semiconductor photo-electrocatalysts.

References

1. Coronado, J.M., A Historical Introduction to Photocatalysis, in: *Design of Advances Photocatalytic Materials for Energy and Environmental Applications*, pp. 1–4, London: Springer London, 2013.
2. Avar, B., Panigrahi, M., Soguksu, A.K., Rajendrachari, S., Gundes, A., Photocatalytic Activity of Soft Magnetic Fe80–xCoxZr10Si10 (x = 0, 40, and 80) Nanocrystalline Melt-Spun Ribbons. *Top. Catal.*, 1–10, 2022.
3. Kiran, K.S., Shashanka, R., Lokesh, S.V., Enhanced Photocatalytic Activity of Hydrothermally Synthesized Perovskite Strontium Titanate Nanocubes. *Top. Catal.*, 1–10, 2022.
4. Pavitra, V., Praveen, B.M., Nagaraju, G., Shashanka, R., Energy Storage, Photocatalytic and Electrochemical Nitrite Sensing of Ultrasound-Assisted Stable Ta2O5 Nanoparticles. *Top. Catal.*, 1–14, 2022.
5. Melchionna, M. and Fornasiero, P., Updates on the Roadmap for Photocatalysis. *ACS Catal.*, **10**, 10, 5493–5501, 2020.
6. Zinola, C.F., Martins, M.E., Tejera, E.P., Neves, N.P., Electrocatalysis: Fundamentals and Applications. *Int. J. Electrochem.*, **2012**, 1–2, 2012.
7. Watzele, S.A., Garlyyev, B., Gubanova, E., Bandarenka, A.S., Structure-reactivity relations in electrocatalysis, in: *Comprehensive Inorganic Chemistry III*, pp. 419–436, Elsevier, 2023.
8. Ameta, R., Solanki, M.S., Benjamin, S., Ameta, S.C., Photocatalysis, in: *Advanced Oxidation Processes for Waste Water Treatment*, pp. 135–175, Academic Press, Elsevier, 2018.
9. Zhu, S. and Wang, D., Photocatalysis: Basic Principles, Diverse Forms of Implementations and Emerging Scientific Opportunities. *Adv. Energy Mater.*, 7, 23, 1700841, 2017.
10. Náfrádi, M., Veréb, G., Firak, D.S., Alapi, T., Photocatalysis: Introduction, Mechanism, and Effective Parameters, *Green Photocatalytic Semiconductors: Recent Advances and Applications*, pp. 3–31, Springer, Cham, 2022.
11. Khan, M.M., Principles and mechanisms of photocatalysis, in: *Photocatalytic Systems by Design*, pp. 1–22, Elsevier, 2021.
12. Zhang, J., Tian, B., Wang, L., Xing, M., Lei, J., Mechanism of Photocatalysis, In: *Photocatalysis. Lecture Notes in Chemistry*, **100**, 1–15, Springer, Singapore, 2018.
13. Fang, Y., Zheng, Y., Fang, T., Chen, Y., Zhu, Y., Liang, Q., Sheng, H., Li, Z., Chen, C., Wang, X., Photocatalysis: an overview of recent developments and technological advancements. *Sci. China Chem.*, **63**, 2, 149–181, 2020.
14. Lad, V.N. and Murthy, Z.V.P., Advanced materials for photocatalytic applications: the challenge ahead, in: *Handbook of Smart Photocatalytic Materials*, pp. 3–8, Elsevier, 2020.
15. Coronado, J.M., Photons, Electrons and Holes: Fundamentals of Photocatalysis with Semiconductors, in: *Design of advanced photocatalytic*

materials for energy and environmental applications, pp. 5–33, London: Springer London, 2013.
16. Wang, L., Zhao, J., Liu, H., Huang, J., Design, modification and application of semiconductor photocatalysts. *J. Taiwan Inst. Chem. Eng.*, **93**, 590–602, 2018.
17. Yang, X. and Wang, D., Photocatalysis: From Fundamental Principles to Materials and Applications. *ACS Appl. Energy Mater.*, **1**, 12, 6657–6693, 2018.
18. Liu, C., Dong, X., Hao, Y., Wang, X., Ma, H., Zhang, X., A novel supramolecular preorganization route for improving $g\text{-}C_3N_4/g\text{-}C_3N_4$ metal-free homojunction photocatalysis. *New J. Chem.*, **41**, 20, 11872–11880, 2017.
19. Ge, S.-C., Te, S., Alsb, S., Gaas, G., Energy band profiles of heterojunctions, in: *Semiconductor Heterojunctions*, pp. 24–37, 2015.
20. Fresno, F., Heterojunctions: Joining Different Semiconductors, In *Design of Advanced Photocatalytic Materials for Energy and Environmental Applications*, pp. 311–327, London: Springer London, 2013.
21. Low, J., Yu, J., Jaroniec, M., Wageh, S., Al-Ghamdi, A.A., Heterojunction Photocatalysts. *Adv. Mater.*, **29**, 20, 1601694, 2017.
22. Lu, L., Wu, B., Shi, W., Cheng, P., Metal–organic framework-derived heterojunctions as nanocatalysts for photocatalytic hydrogen production. *Inorg. Chem. Front.*, **6**, 12, 3456–3467, 2019.
23. Xu, Q., Zhang, L., Yu, J., Wageh, S., Al-Ghamdi, A.A., Jaroniec, M., Direct Z-scheme photocatalysts: Principles, synthesis, and applications. *Mater. Today*, **21**, 10, 1042–1063, 2018.
24. Xu, Q., Zhang, L., Cheng, B., Fan, J., Yu, J., S-Scheme Heterojunction Photocatalyst. *Chem*, **6**, 7, 1543–1559, 2020.
25. Hagen, J., Electrocatalytic Processes, in: *Industrial Catalysis: A Practical Approach*, pp. 319–333, Weinheim, Germany: Wiley-VCH, 2015.
26. Maheshwari, S., Li, Y., Agrawal, N., Janik, M.J., Density functional theory models for electrocatalytic reactions. *Adv. Catal.*, 63, 117–167, 2018. Academic Press Inc.
27. Dyantyi, N., Calderón Gómez, J.C., Mekuto, L., Bujlo, P., Pattrick, G., Electrocatalysts: selectivity and utilization, in: *PEM Fuel Cells*, pp. 55–70, Elsevier, 2022.
28. Chen, W., Liu, S., Fu, Y., Yan, H., Qin, L., Lai, C., Zhang, C., Ye, H., Chen, W., Qin, F., Xu, F., Huo, X., Qin, H., Recent advances in photoelectrocatalysis for environmental applications: Sensing, pollutants removal and microbial inactivation. *Coord. Chem. Rev.*, **454**, 214341, 2022.
29. Lewerenz, H.J., Heine, C., Skorupska, K., Szabo, N., Hannappel, T., Vo-Dinh, T., Campbell, S.A., Klemm, H.W., Muñoz, A.G., Photoelectrocatalysis: principles, nanoemitter applications and routes to bio-inspired systems. *Energy Environ. Sci.*, **3**, 6, 748, 2010.
30. Bessegato, G.G., Guaraldo, T.T., de Brito, J.F., Brugnera, M.F., Zanoni, M.V.B., Achievements and Trends in Photoelectrocatalysis: from Environmental to Energy Applications. *Electrocatalysis*, **6**, 5, 415–441, 2015.

31. Yang, J., Wang, D., Han, H., Li, C., Roles of Cocatalysts in Photocatalysis and Photoelectrocatalysis. *Acc. Chem. Res.*, **46**, 8, 1900–1909, 2013.
32. Bessegato, G.G., Guaraldo, T.T., Zanoni, M.V.B., Enhancement of Photoelectrocatalysis Efficiency by Using Nanostructured Electrodes, in: *Modern Electrochemical Methods in Nano, Surface and Corrosion Science*, Janeza Trdine 9,51000 Rijeka, Croatia: InTech, 2014.
33. Khan, S.U.M., Materials for photoelectrochemical devices, in: *Materials for Energy Conversion Devices: A Volume in Woodhead Publishing Series in Electronic and Optical Materials*, pp. 35–62, Elsevier Inc, 2005.
34. Cao, D., Wang, Y., Zhao, X., Combination of photocatalytic and electrochemical degradation of organic pollutants from water. *Curr. Opin. Green Sustain. Chem.*, **6**, 78–84, 2017.
35. Yang, X., Wolcott, A., Wang, G., Sobo, A., Fitzmorris, R.C., Qian, F., Zhang, J.Z., Li, Y., Nitrogen-Doped ZnO Nanowire Arrays for Photoelectrochemical Water Splitting. *Nano Lett.*, **9**, 6, 2331–2336, 2009.
36. Zheng, Y.-Z., Zhao, J., Zhang, H., Chen, J.-F., Zhou, W., Tao, X., Dual-functional ZnO nanorod aggregates as scattering layer in the photoanode for dye-sensitized solar cells. *Chem. Commun.*, **47**, 41, 11519, 2011.
37. Lin, C.-Y., Lai, Y.-H., Chen, H.-W., Chen, J.-G., Kung, C.-W., Vittal, R., Ho, K.-C., Highly efficient dye-sensitized solar cell with a ZnO nanosheet-based photoanode. *Energy Environ. Sci.*, **4**, 9, 3448, 2011.
38. Soni, I., Kumar, P., Kudur Jayaprakash, G., Recent advancements in the synthesis and electrocatalytic activity of two-dimensional metal–organic framework with bimetallic nodes for energy-related applications. *Coord. Chem. Rev.*, **472**, 214782, 2022.
39. Wang, Q., Gao, Q., Al-Enizi, A.M., Nafady, A., Ma, S., Recent advances in MOF-based photocatalysis: environmental remediation under visible light. *Inorg. Chem. Front.*, **7**, 2, 300–339, 2020.
40. Meng, S., Li, G., Wang, P., He, M., Sun, X., Li, Z., Rare earth-based MOFs for photo/electrocatalysis. *Mater. Chem. Front.*, **7**, 5, 806–827, 2023.
41. Candia-Onfray, C., Rojas, S., Zanoni, M.V.B., Salazar, R., An updated review of metal–organic framework materials in photo(electro)catalytic applications: From CO2 reduction to wastewater treatments. *Curr. Opin. Electrochem.*, **26**, 100669, 2021.
42. Xiao, J.-D. and Jiang, H.-L., Metal–Organic Frameworks for Photocatalysis and Photothermal Catalysis. *Acc. Chem. Res.*, **52**, 2, 356–366, 2019.
43. Ali, M., Pervaiz, E., Noor, T., Rabi, O., Zahra, R., Yang, M., Recent advancements in MOF- based catalysts for applications in electrochemical and photoelectrochemical water splitting: A review. *Int. J. Energy Res.*, **45**, 2, 1190–1226, 2021.
44. Guo, K., Hussain, I., Jie, G., Fu, Y., Zhang, F., Zhu, W., Strategies for improving the photocatalytic performance of metal-organic frameworks for CO2 reduction: A review. *J. Environ. Sci.*, **125**, 290–308, 2023.

45. Alulema-Pullupaxi, P., Espinoza-Montero, P.J., Sigcha-Pallo, C., Vargas, R., Fernández, L., Peralta-Hernández, J.M., Paz, J.L., Fundamentals and applications of photoelectrocatalysis as an efficient process to remove pollutants from water: A review. *Chemosphere*, **281**, 130821, 2021.
46. Mizrahi, A. and Meyerstein, D., Plausible roles of carbonate in catalytic water oxidation, in: *Advances in Inorganic Chemistry*, **74**, 343–360, 2019.
47. Li, R. and Li, C., Photocatalytic Water Splitting on Semiconductor-Based Photocatalysts, in: *Advances in catalysis*, **60**, 1–57, 2017.
48. Peng, M., Guan, G., Deng, H., Han, B., Tian, C., Zhuang, J., Xu, Y., Liu, W., Lin, Z., PCN-224/rGO nanocomposite based photoelectrochemical sensor with intrinsic recognition ability for efficient p-arsanilic acid detection. *Environ. Sci. Nano*, **6**, 1, 207–215, 2019.
49. Cao, Y., Wang, L., Wang, C., Hu, X., Liu, Y., Wang, G., Sensitive detection of glyphosate based on a Cu-BTC MOF/g-C3N4 nanosheet photoelectrochemical sensor. *Electrochim. Acta*, **317**, 341–347, 2019.
50. Jia, M., Yang, Z., Xu, H., Song, P., Xiong, W., Cao, J., Zhang, Y., Xiang, Y., Hu, J., Zhou, C., Yang, Y., Wang, W., Integrating N and F co-doped TiO2 nanotubes with ZIF-8 as photoelectrode for enhanced photo-electrocatalytic degradation of sulfamethazine. *Chem. Eng. J.*, **388**, 124388, 2020.
51. Wang, J., Xue, C., Yao, W., Liu, J., Gao, X., Zong, R., Yang, Z., Jin, W., Tao, D., MOF-derived hollow TiO2@C/FeTiO3 nanoparticles as photoanodes with enhanced full spectrum light PEC activities. *Appl. Catal. B*, **250**, 369–381, 2019.
52. Lu, H., Wang, Z., Wang, L., Photocatalytic and Photoelectrochemical Carbon Dioxide Reductions toward Value-Added Multicarbon Products. *ACS ES&T Eng.*, **2**, 6, 975–988, 2022.
53. Majumdar, P., Bera, M.K., Pant, D., Patra, S., Enzymatic Electrocatalysis of CO2 Reduction, in: *Encyclopedia of Interfacial Chemistry*, pp. 577–589, Elsevier, 2018.
54. Jiao, X., Zheng, K., Hu, Z., Sun, Y., Xie, Y., Broad-Spectral-Response Photocatalysts for CO2 Reduction. *ACS Cent. Sci.*, **6**, 5, 653–660, 2020.
55. Fu, Z., Yang, Q., Liu, Z., Chen, F., Yao, F., Xie, T., Zhong, Y., Wang, D., Li, J., Li, X., Zeng, G., Photocatalytic conversion of carbon dioxide: From products to design the catalysts. *J. CO2 Util.*, **34**, 63–73, 2019.
56. Ochedi, F.O., Liu, D., Yu, J., Hussain, A., Liu, Y., Photocatalytic, electrocatalytic and photoelectrocatalytic conversion of carbon dioxide: a review. *Environ. Chem. Lett.*, **19**, 2, 941–967, 2021.
57. Gulati, S., Vijayan, S., Mansi, Kumar, S., Harikumar, B., Trivedi, M., Varma, R.S., Recent advances in the application of metal-organic frameworks (MOFs)-based nanocatalysts for direct conversion of carbon dioxide (CO2) to value-added chemicals. *Coord. Chem. Rev.*, **474**, 214853, 2023.
58. Chen, P., Zhang, Y., Zhou, Y., Dong, F., Photoelectrocatalytic carbon dioxide reduction: Fundamental, advances and challenges. *Nano Mater. Sci.*, **3**, 4, 344–367, 2021.

59. Cardoso, J.C., Stulp, S., de Souza, M.K.R., Hudari, F.F., Gubiani, J.R., Frem, R.C.G., Zanoni, M.V.B., The effective role of ascorbic acid in the photoelectrocatalytic reduction of CO2 preconcentrated on TiO2 nanotubes modified by ZIF-8. *J. Electroanal. Chem.*, **856**, 113384, 2020.
60. Chen, S., Hai, G., Gao, H., Chen, X., Li, A., Zhang, X., Dong, W., Modulation of the charge transfer behavior of Ni(II)-doped NH2-MIL-125(Ti): Regulation of Ni ions content and enhanced photocatalytic CO2 reduction performance. *Chem. Eng. J.*, **406**, 126886, 2021.
61. Silva, B.C., Irikura, K., Flor, J.B.S., dos Santos, R.M.M., Lachgar, A., Frem, R.C.G., Zanoni, M.V.B., Electrochemical preparation of Cu/Cu2O-Cu(BDC) metal-organic framework electrodes for photoelectrocatalytic reduction of CO2. *J. CO2 Util.*, **42**, 101299, 2020.
62. Cardoso, J.C., Stulp, S., de Brito, J.F., Flor, J.B.S., Frem, R.C.G., Zanoni, M.V.B., MOFs based on ZIF-8 deposited on TiO2 nanotubes increase the surface adsorption of CO2 and its photoelectrocatalytic reduction to alcohols in aqueous media. *Appl. Catal. B*, **225**, 563–573, 2018.
63. Irikura, K., Perini, J.A.L., Flor, J.B.S., Frem, R.C.G., Zanoni, M.V.B., Direct synthesis of Ru3(BTC)2 metal-organic framework on a Ti/TiO2NT platform for improved performance in the photoelectroreduction of CO2. *J. CO2 Util.*, **43**, 101364, 2021.
64. Qian, W., Xu, S., Zhang, X., Li, C., Yang, W., Bowen, C.R., Yang, Y., Differences and Similarities of Photocatalysis and Electrocatalysis in Two-Dimensional Nanomaterials: Strategies, Traps, Applications and Challenges. *Nano Micro Lett.*, **13**, 1, 156, 2021.
65. Tay, N.E.S., Lehnherr, D., Rovis, T., Photons or Electrons? A Critical Comparison of Electrochemistry and Photoredox Catalysis for Organic Synthesis. *Chem. Rev.*, **122**, 2, 2487–2649, 2022.
66. Wu, H., Tan, H.L., Toe, C.Y., Scott, J., Wang, L., Amal, R., Ng, Y.H., Photocatalytic and Photoelectrochemical Systems: Similarities and Differences. *Adv. Mater.*, **32**, 18, 1904717, 2020.
67. Chen, A., Electrocatalysis and photoelectrochemistry based on functional nanomaterials. *Can. J. Chem.*, **92**, 7, 581–597, 2014.
68. Marinho, B.A., Suhadolnik, L., Likozar, B., Huš, M., Marinko, Ž., Čeh, M., Photocatalytic, electrocatalytic and photoelectrocatalytic degradation of pharmaceuticals in aqueous media: Analytical methods, mechanisms, simulations, catalysts and reactors. *J. Clean. Prod.*, **343**, 131061, 2022.

5

A Review on Photocatalytic Fuel Cell (PFC) for Sustainable Wastewater Treatment and Renewable Energy Production

Soumasree Chatterjee[1]* and Elvis Fosso-Kankeu[2]

[1]Department of Civil Engineering, Indian Institute of Technology, Kanpur, India
[2]Department of Metallurgy, Faculty of Engineering and Built Environment, University of Johannesburg, Doornfontein, Johannesburg, South Africa

Abstract

The environmental pollution due to the rapid discharge of effluent is increasing exponentially in the last few decades. Several conventional methods, such as reverse osmosis, chemical precipitation, adsorption, chemical oxidation, and membrane filtration, have been applied for wastewater treatment. Nonetheless, they have their own limitations. Concurrently, fossil fuels have been extensively used as a source of energy that has led to world critical energy shortage and several environmental problems, such as increased greenhouse gases, global heating, climate change, environmental pollution, and health problems. Renewable energy production is now attractive for a host of researchers because of its environment friendly benign nature. Therefore, wastewater treatment and energy scarcity are two main issues from the last few decades. Thus, an environment-friendly approach having suitable option of both wastewater management and clean energy production is in search. Photocatalytic oxidation has been considered to degrade the organic pollutants present in air and water. Recently, photocatalysis (PC) is assembled with fuel cell (FC) to develop photocatalytic fuel cell (PFC) for dual purposes including wastewater treatment and renewable energy production. PFC system, a green technology, can degrade an extensive range of pollutants, including phenols, drugs, and dyes (*viz.*, methylene green, methylene blue, azo dye, Rhodamine B, Congo red, and methyl orange). The organic compounds

**Corresponding author*: soumasree.chatterjee@gmail.com

Sadanand Pandey, Elvis Fosso-Kankeu and Soumya Pandit (eds.) Synergy of Bio-Chemical Processes for Photocatalytic and Photoelectrochemical Wastewater Treatment, (103–138) © 2024 Scrivener Publishing LLC

present in industrial and domestic wastewaters can be considered as the sources of chemical energy. The efficiency of PFC system depends on several features including electrode materials, open-circuit voltage (V_{OC}), short-circuit current (Jsc), fill factor (FF), light intensity, pH of initial solution, nature and concentration of electrolyte, and molecular structure of organic pollutant. The main purpose of PFC is to recover and alter the chemical energy present in the organic pollutants into power energy instead of consuming electricity during wastewater treatment. The present chapter covers the principle and design of PFC systems including different electrodes, degradation of numerous organic pollutants present in wastewater, and simultaneous renewable energy production.

Keywords: Photocatalytic fuel cell, photocatalysis, wastewater, pollutant degradation, energy production

5.1 Introduction

Nowadays, environmental pollution is a matter of great concern and considered as a global crisis. Due to rapid industrial growth, a huge amount of wastewater-containing organic pollutants, such as aromatic hydrocarbons, dyes, pharmaceuticals, and other persistent organic complexes, are coming into the water body, which have a negative impact on human health and environment [1]. Organic pollutants cause marine pollution and require energy for the purification of contaminated water. Organic compounds or pollutants present in water and wastewater are important source of chemically stored energy [2–10]. Nonetheless, it is problematic to recover the energy using traditional wastewater treatment methods [11]. Various traditional wastewater treatment processes, including chemical precipitation and oxidation, biological treatment, adsorption, membrane separation, and electrochemical technology, basically emphasize on the removal of waste instead of energy recovery that associates with higher operation cost [1]. There are also some other methods for wastewater treatment, *viz.*, reverse osmosis (RO), that use huge amount of energy for, e.g., 5–100 kwh/kg chemical oxygen demand (COD) [11, 12]. An effective way to overwhelm this problem is by adding a component of advanced oxidation processes (AOPs) with the traditional water treatment device [12, 13]. Advanced oxidation processes (AOPs), for example, UV ozonation, photocatalysis, and UV-peroxidation with hydrogen peroxide, are very efficient photochemical technologies to decompose an extensive of organic compounds and produce reactive chemicals [11, 13]. Among them, semiconductor-based photocatalytic techniques are considered as most suitable approach due to their lower toxicity, chemical inertness, higher stability, lower cost, and better efficiency [12]. However, in the case of AOPs, the lower level of chemical energy was achieved during the decomposition of organic compounds that is the major drawback of this process [12, 14]. In the last few years, a combined biological and electrochemical process called microbial fuel cells (MFCs) was fabricated for the recovery of energy from waste [12, 15]. The main advantage of MFCs is that it can utilize a wide variety of substrate such as phenol, glucose, and ethanol present in wastewater and generate electricity [15, 16]. However, there are some limitations of MFCs such as difficulty in bacterial cultivation, lower bio-electrochemical activity, lower microbial selectivity toward organic compounds, and prolonged start-up time [17]. Thus, an efficient organic pollutant removal technology with consecutive recovery of chemical energy is an urgent necessity.

In contrast to pre-existing traditional MFC system, photocatalytic fuel cell (PFC) has been found to be more efficient [16–30]. This technique has dual advantages similar to MFCs, such as decomposing of organic pollutants through photocatalysis process and concurrently generation of electricity instead of consuming electricity. Compared to traditional MFCs, PFCs have several benefits including greater efficiency, simplicity, removal of toxicity, and ability to degrade an extensive range of compounds including heavy metal pollutants, dyes, alcoholic compounds, and antibiotics [13]. PFCs are more appropriate for the treatment of wastewater and concurrently produce renewable energy in remote and poor regions, as it has the ability to use sunlight that is easily available and free of cost [30].

Several research works have been concentrated on the treatment of different organic pollutants and synchronized electricity generation via PFC [16–30]. Although, a very few review articles about PCFs are available in the literature [11–13, 17, 31]. Herein, the focus is on the application of PFCs for sustainable wastewater treatment and simultaneous renewable energy production considering degradation of various organic pollutants, different electrodes, and electricity generation, as it is a great interest in the domain of photocatalytic technology, environmental science and engineering, chemistry, clean and green technology, etc.

5.2 Photocatalysis

5.2.1 Theory

The word "photocatalysis" is the combination of two terms: photo that is associated to photon and catalysis that is a process that alters the rate of chemical reaction in the existence of catalyst. Photocatalysis is a phenomenon that alters the rate of chemical reaction using light energy in the presence of catalyst [13, 32]. Photocatalysts are generally semiconductors in nature. Chlorophyll present in plants is a natural example of photocatalyst. This process oxidizes organic compounds into simpler form and is extensively used in the decomposition of organic pollutants present in water and wastewater, purification of air, production of hydrogen, and antibacterial activity. Various types of photocatalysts, including bismuth vanadate ($BiVO_4$), zinc oxide (ZnO), titanium dioxide (TiO_2), and tungsten trioxide (WO_3), are synthesized and employed for the degradation of organic pollutants [17]. Among them, TiO_2-based nano semiconductor is identified as more competent nanostructured photocatalyst [17]. Nowadays, photocatalysis, a green technology, is more attractive because of several benefits including cost effectiveness and higher efficiency [33].

5.2.2 Mechanism

According to energy band theory of solids [17], the difference in energy (in terms of eV) between the bottom of the conduction band (CB) and the top of the valence band (VB) is known as bandgap energy (E_g) [17]. When the energy of irradiated light is higher than or equal to E_g of the photocatalyst, VB becomes excited, and excess energy of the electron stimulates the electron to CB. Therefore, the oxidation-reduction reaction happens on VB and CB, respectively [13, 17, 32]. Reaction mechanism of photocatalysis is mentioned below (Equations 5.1 to 5.3):

Photocatalyst + light energy (hY) → Photocatalyst (e^- present in conduction band + h^+ present in valence band)

(5.1)
[34]

Photocatalyst (h^+_{VB}) + H_2O (oxidizes) → Photocatalyst + H^+ (hydrogen ion) + $\cdot OH$ (hydroxyl radical)

(5.2)
[34]

Photocatalyst (h^+_{VB}) + OH^- (hydroxyl ion) (adsorption) → Photocatalyst + $\cdot OH$ (hydroxyl radical)

(5.3)
[34]

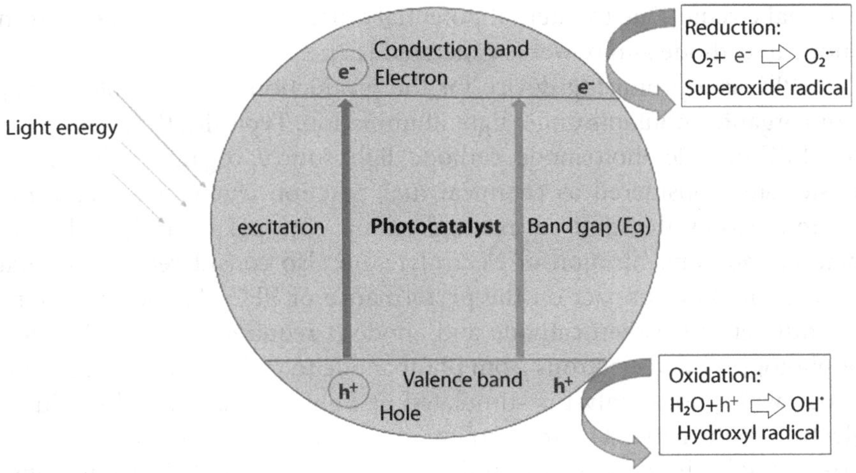

Figure 5.1 Representation of photocatalysis [35].

As per literature, there are basically three types of photocatalytic mechanism: (i) irradiation of light stimulates the photoexcited electron (e^-) and promotes to the CB of the photocatalyst, although the photoexcited hole (h^+) stays on the VB [17]; (ii) photoexcited charge localizes to the surface of the photocatalyst for redox reaction [17]; and (iii) photoexcited electrons interact with both hydrogen ion (H^+) in the absence of oxygen to generate H_2 and dissolved O_2 in the electrolyte to obtain superoxide radicals ($\cdot O_2^-$) [17]. The photoexcited hole present on the VB interacts with H_2O or OH^- and hydroxyl radical ($\cdot OH$) can be generated [17]. The organic pollutants can be degraded to simpler form by both $\cdot OH$ and $\cdot O_2^-$ because of their oxidizing property in nature. Figure 5.1 represents an overview diagram of photocatalysis.

5.3 Photocatalytic Fuel Cell

5.3.1 Theory

Photocatalysis technique has been assembled with fuel cell operation called PFC. A fuel cell is an apparatus that has the ability to convert chemical energy present in the fuel, including methanol, hydrogen, and ethanol into power energy via electrochemical process [17]. This technique is cost-effective and eco-friendly in nature. Usually, a fuel cell comprises of a cathode and an anode, in which they are linked by chemical fuel and an external circuit. Due to internal potential difference, electrons move from anode to cathode and power is generated [17].

A PFC is a promising device for the production of renewable energy from organic pollutants under light illumination. Typically, the major parts of a PFC include photoanode, cathode, light source, organic pollutants in wastewater considered as chemical fuel, reaction chamber for carrying organic wastewater, and an external circuit. The pH of initial solution, nature, and concentration of electrolyte are also considered as the vital factors that have impact on the performance of PFC. An electro potential difference between cathode and anode is required for the migration of photogenerated electrons from photoanode to cathode. In PFC, photoanode with photocatalyst is stimulated by light irradiation and produces electron–hole pairs. Organic pollutants are oxidized and degraded by positive holes in photoanode, and, then, photogenerated electrons transfer toward cathode through the external circuit and eventually generate electricity [1, 17].

5.3.2 Features

The following are some features of PFC system:

- PFC has the ability to degrade the organic pollutants present in the water and wastewater under light irradiation.
- Fabrication and operation of PFC system is easier than traditional removal methods. The higher performance is also obtained from PFC.
- Type of electrodes, pH of initial solution, nature and concentration of electrolyte, molecular structure of organic pollutants, etc., are considered as the major factors that have immense impact on efficiency of PFC.
- PFC can degrade organic pollutants that cannot be efficiently decomposed by microorganism.
- PFC is capable to degrade an extensive range of pollutants due to oxidative property of photogenerated holes (h^+) and hydroxyl radicals (•OH).
- Electrode materials can now be modified and synthesized using nanotechnology for better performance.

The oxidation process is influenced by chemical composition and structure of the substrate having different groups, such as carboxylic acids, alcohols, aromatics, amines, and saccharides [12]. PFC efficacy is reduced when the substrate concentration is less than the optimal value. However, the rate of the chemical reaction is no longer restricted by mass transport in the presence of higher amount of substrate [12, 17].

A definite amount of electrolyte, such as nitrates, sulfates, phosphates, alkali (*viz.*, NaOH and KOH), sodium chloride, and persulfate salts, is typically applied to the reaction chamber to increase the charge movement and to decrease the internal resistance of the fuel cell, which leads to enhance the photochemical performance and to produce higher energy in output [12]. Among the electrolytes, sulfate (Na_2SO_4) is mostly considered by the researchers, due to its stable and inert chemical properties [12]. Various organic compounds or pollutants such as phenols, drugs, and dyes, *viz.*, methylene green, methylene blue, azo dye, Rhodamine B, Bisphenol A, Reactive black 5, Congo red, and methyl orange, can work as fuel in PFC [16, 36–38, 30].

5.3.3 Working Principle

A photoanode with photocatalyst produces electrons (e^-) to decompose organic compounds and to produce electricity under light illumination. A diagram of PFC system is presented in Figure 5.2. The efficiency of PFC is affected by various features, such as the type and setup of electrodes and the construction of the system. Different reactions take place at photoanode and cathode of PFC system, and intermediates are mentioned below [11, 13, 39]:

Step 1: Photoexcitation occurred at photoanode carrying photocatalyst under light irradiation

Photoanode with photocatalyst + light irradiation → h^+ (valence band) + e^- (conduction band)

(5.4)
[13, 11]

Step 2: Interaction with H_2O and production of hydroxyl radicals

h^+ (valence band) + H_2O → hydroxyl radicals (•OH) + H^+ (5.5)
[13, 11]

Step 3: Redox reaction process
(i) Oxidation process with organic pollutants

$$h^+ \text{ (valence band)} + \text{organic pollutant} \rightarrow \text{organic pollutant}^{\cdot +} \quad (5.6)$$
[13, 11]

(ii) Decomposition of organic pollutants

$$\begin{aligned} &\text{hydroxyl radicals } (''\text{OH}) + \text{organic pollutant} \rightarrow \text{organic pollutant}^{\cdot} \\ &\text{organic pollutant}^{\cdot} + O_2 \rightarrow \text{organic pollutant - OO}^{\cdot} \\ &\text{superoxide radicals } (O_{2^{\cdot}}) + \text{organic pollutant}^{\blacklozenge} \rightarrow \text{organic pollutant-OO}^{\cdot} \end{aligned} \quad (5.7)$$
[13, 11]

(iii) Reduction process with organic pollutants

$$e^- + \text{organic pollutant} \rightarrow \text{organic pollutants} \quad (5.8)$$
[13, 11]

(iv) Migration of electron, production of H_2O and hydrogen peroxide

$$\begin{aligned} &\text{Organic pollutant}^{\blacklozenge} \rightarrow \text{organic pollutant}^{++} + e^- (\text{conduction band}) \\ &4H^+ \text{ ion} + O_2 + 4e^- \rightarrow 2H_2O \\ &2H_2O + 2H+ \text{ ion} + 2e^- \rightarrow \text{hydrogen peroxide } (H_2O_2) \end{aligned} \quad (5.9)$$
[13, 11]

Step 4: Production of superoxide radicals at cathode

$$e^- \text{ (conduction band)} + O_2 \rightarrow \text{superoxide radicals } (O_2^-) \quad (5.10)$$
[13, 11]

At first, the photoexcitation process takes place at the photoanode, carrying photocatalysts under light irradiation. The excitation energy is equivalent to or higher than photocatalyst energy band [11]. Electrons (e^-) are excited at CB, and holes (h^+) remain in the VB (Equation 5.4) [11].

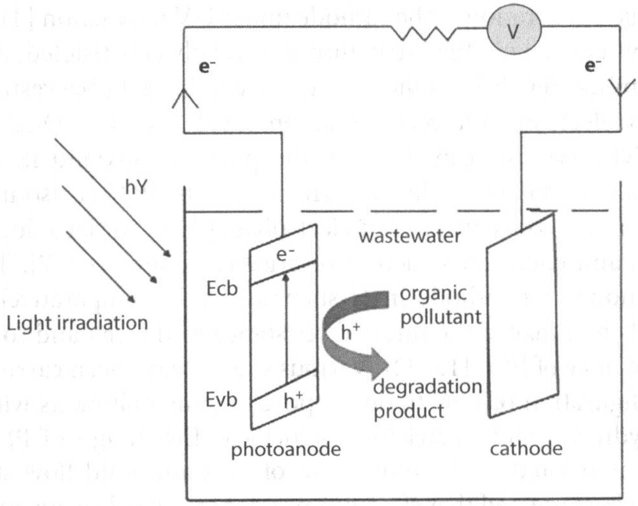

Figure 5.2 Schematic diagram of PFC system [1].

The photogenerated electrons and holes are separated from anode (Equation 5.5) [11]. Due to decomposition of organic pollutants at anode, electrons migrate toward cathode via an external circuit (Equation 5.6) [11]. Hydrogen ions move toward the cathode by diffusion. At the cathode, electrons interact with oxygen and produce H_2O or superoxide radicals (Equations 5.7 to 5.10) [11].

The power density in PFC system can be estimated using the following equation [31]:

$$\text{Power density} = \text{voltage} \times \text{current density}$$

where power density is measured in mW/cm^2, current density is measured in mA/cm^2, and voltage is measured in mV.

5.3.4 Design

As per the literature, hydrogen gas was first generated by water reduction techniques via photoelectrochemical cell carrying a single photocatalyst. Fujishima and Honda (1972) [40] developed the water splitting by photocatalysis via single crystalline made of TiO2 electrode. Afterward, Kaneko *et al.* (2005) [41] first addressed about PFC. They employed TiO_2-based FTO glass, Pt, and NH_3 as photoanode, cathode, and degrading substance, respectively, and observed the photodegradation of NH_3 at photoanode and hydrogen generation at the cathode under UV irradiation [41]. Single-chamber photocatalytic PFC structure was usually constructed. Although, single-chamber PFC is less efficient due to smaller light harvesting ability and higher electron–hole recombination capability [17]. Dual-chamber photocatalytic systems can decrease the polarizations and increase the performance of PFC [17]. The performance of the PFC is also influenced by selection and concentration of electrolyte, pH of initial solution, concentration, and chemical structure of organic pollutants [17]. The space between anode and cathode in H-shaped PFC is comparatively higher, which leads to enhance the internal resistance of the cell and to decrease the performance of PFC [12, 42]. Various studies have been carried out for better configuration of PFC to decompose organic pollutants with simultaneous hydrogen and electricity production. The design of PFC device is directly associated to the movement of ions and fluid flow state [12]. Few researchers noticed that electrons present in cathode were not entirely used in oxygen reduction process [43–46]. Therefore, few studies have been emphasized on using the electrons for CO_2 reduction [47], hydrogen evolution reaction [48], and H_2O_2 generation [49] to obtain green fuels via the cathode of PFC. In this present chapter, two types of PFC, such as single-photoelectrode and dual-photoelectrode PFCs, are discussed.

5.3.4.1 *Single-Photoelectrode PFC*

Typically, a single-photoelectrode PFC consists of a photoanode with semiconductor, a cathode, electrolytes, substrates, and light source. Both electrodes are linked by an external circuit. Due to an electro potential changes between anode and cathode, the photogenerated electrons move from photoanode to cathode under light irradiation. The potential difference between anode and cathode is evaluated by the Fermi level of the photocatalyst and reduction reaction, respectively [12]. There are basically two types of semiconductors such as n-type and p-type. The Fermi level defines as the thermodynamic effort needed for accumulation of one electron to the solid material. As per literature, the Fermi levels of n-type and p-type semiconductors are present near the conduction band and valance band, respectively [12, 17, 19, 50]. Typically, photoanodes employed in PFC are TiO_2, ZnO, WO_3, α-Fe_2O_3, $BaTiO_3$, $BiVO_4$, etc. [2, 3, 6–8]. However, photoanodes made of TiO_2 and non–photo-responsive commercial platinum/Pt-black cathode are mostly applied in photoelectrode PFC. Application of TiO_2 as a photoanode is increased because of its stable physical and chemical characteristics, non-toxicity, low price, simple method of preparation, and optoelectronic nature [31]. The radical reactions take place on the surface area of photoanode and cathode. Photoexcited electrons move from anode to cathode that leads to energy recovery from organic pollutants. Liu *et al.* (2011) [5] developed a PFC for power production from several organic compounds and simultaneous wastewater treatment using TiO_2-based nanotube array. This system responded to visible light and produced higher cell efficiency by combining TiO_2 nanotubes with CdS or Cu_2O [5]. Xie *et al.* (2018) [8] developed a PFC system induced by visible light where cathode was made of platinum/C and photoanode was synthesized from $BiVO_4/WO_3$. Maximum power density (P_{max}) of 8.58 μW/cm^2 and short-circuit current density (I_{sc}) of 91.8 μA/cm^2 were obtained when glucose was used [8]. They also reported that the decomposition rate of tetracycline hydrochloride using this type of PFC was about 87.2% within 8 h that was higher than other processes. Chen *et al.* (2020) [2] observed the decomposition of ofloxacin (OFX) in water using PFC made of UVA-LED/TiO_2 and also studied the effect of pH and dissolved oxygen (DO) on decomposition.

However, there are few weaknesses in this system. Pt cathode utilized in this system is costly and rare that leads to problem in large-scale operation. The photo-simulated surface area of photoanode is restricted by huge particle size, which delays the photocatalytic process. Therefore, the efficiency of single-photoelectrode PFC is much lower than expected value [12, 17]. Currently, to reduce the operational cost and enhance the pollutants degradation performance, different types of cathodes are fabricated with Pt-free, biocathode, etc. [12, 51–53].

5.3.4.2 *Dual-Photoelectrode PFC*

In case of dual-photoelectrode PFC, photoanode and photocathode are usually made of n-type and p-type semiconductors, respectively, and solar light irradiation can be used as light source. The main factor to construct this PFC is the selection of electrode materials. Both electrodes are excited under light illumination and show an opposite inner bias that stimulates the electron to flow through the external circuit. Compared to single-photoelectrode PFC, the dual-photoelectrode PFC shows some benefits, such as higher stability, improved photoexcited ion migration rate, greater harvesting of light, and higher efficacy to transform organics to energy [12, 17, 36, 54]. Chen *et al.* (2012) [37] first informed about a PFC for organic complex removal as well as electricity production that worked in visible light where photoanode was made of WO_3/W and photocathode was developed from Cu_2O/Cu. Li *et al.* (2013) [16] designed a photoanode

Table 5.1 List of different types of PFCs.

Type of PFC	Electrodes (Anode–Cathode)	Organic pollutants	Light source	Electrolyte	Decomposition efficiency (%)	References
Single photoelectrode	TiO2/Ti–Pt	Rhodamine B	UV irradiation	0.05 mol/L Na_2SO_4	90.23	[7]
Single photoelectrode	ZnO–Pt/C	Reactive black 5	UV irradiation	0.1 mol/L NaCl	100	[4]
Single photoelectrode	STNA–Pt Black/Pt	Acetic acid	UV irradiation	0.1 mol/L Na_2SO_4	81	[6]
Single photoelectrode	TNA/Ti–Pt	Ofloxacin	UV irradiation	0.01 mol/L Na_2SO_4	96.8	[2]
Single photoelectrode	AgI/BiOI–Pt	Formic acid	Visible light	0.05 mol/L Na_2SO_4	90.8	[19]
Single photoelectrode	Fe_2O_3/NiO–Pt	-	Visible light	1 mol/L methanol with Na_2SO_4	-	[9]
Single photoelectrode	STNA–Pt-black/Pt	Acetic acid	UV irradiation	0.1 mol/L Na_2SO_4	35	[24]

(Continued)

Table 5.1 List of different types of PFCs. (*Continued*)

Type of PFC	Electrodes (Anode–Cathode)	Organic pollutants	Light source	Electrolyte	Decomposition efficiency (%)	References
Single photoelectrode	$BiVO_4/WO_3$–Pt/C	Tetracycline hydrochloride	Visible-light irradiation	0.1 mol/L KH_2PO_4	87.2	[8]
Single photoelectrode	CdS/TiO_2–Pt	-	UV irradiation	1 mol/L Na_2S	-	[5]
Dual photoelectrode	g-C_3N_4/Fe^0(1%)/ TiO_2–WO_3	Berberine chloride	Visible-light irradiation	0.05 M Na_2SO_4	91	[56]
Dual photoelectrode	$BiVO_4/TiO_2$–ZnO/ CuO	Methylene blue, Methyl organe, and Congo red	Solar-light irradiation	0.1 M Na_2SO_4	83, 76, and 90	[36]
Dual photoelectrode	PANI/TiO_2 NTs–CuO/Co_3O_4 NRs	Rhodamine B	Visible-light irradiation	0.1 mol/L Na_2SO_4	68	[38]
Dual photoelectrode	TiO2/WO3/W–Pt/ BJS	Rhodamine B, Methylene blue, Phenol, Bisphenol A, Dichlorophen, and Tetracycline HCl	Sunlight illumination	0.1 M K_2SO_4	99.3, 98.6, 91.7, 95.2, 96.5, and 92.3	[30]

(*Continued*)

Table 5.1 List of different types of PFCs. (*Continued*)

Type of PFC	Electrodes (Anode–Cathode)	Organic pollutants	Light source	Electrolyte	Decomposition efficiency (%)	References
Dual photoelectrode	$BiVO_4/WO_3/W$–Pt/BJS	Tetracycline HCl	Solar light irradiation	0.1 mol/L sodium sulfate	78	[54]
Dual photoelectrode	WO_3/W- (anode)–Cu_2O/Cu (cathode)	Rhodamine B, Phenol, and Congo red	AM1.5 illumination	0.1 M KH_2PO_4	58, 66, and 73	[37]
Dual photoelectrode	TiO_2/Ti–Cu_2O/Cu	Methylene blue, Methyl orange, and Congo red	AM1.5 solar light irradiation	0.1 M PBS	88, 67, and 63	[16]

made of TiO_2/Ti and photocathode synthesized with Cu_2O/Cu that generated V_{OC} of 0.49 V, a J_{SC} of 0.23 mA/cm^{-2}, and $(JV)_{max}$ of 0.36 mW/cm^{-2} where 0.05 mol/L phenol was employed as substrate. Wu *et al.* (2015) [55] developed a solar-induced dual photoelectrode with PFC composed of TiO_2 NRs/FTO as photoanode and C/Cu NWAs/Cu as photocathode for hydrogen production and wastewater treatment. Xia *et al.* (2016) [54] reported that a photocurrent of 2.01 mA/cm^2 at 0.6V was obtained using $BiVO_4/WO_3$/W-based photoanode where 0.1 M KH_2PO_4 (pH 7) was used as electrolyte. Subsequently, several studies were carried out on dual-photoelectrode PFCs. These results proposed that the open-circuit voltage of PFC system (V_{OC}) mainly govern by the Fermi level difference between the two photoelectrodes. The higher value of V_{OC} is attained at greater Fermi level changes between the photoelectrodes [12]. V_{OC} also improved with the ratio of crystal facet [12]. Table 5.1 summarizes different types of single and dual photoelectrodes used in PFCs.

5.3.5 Electrodes

5.3.5.1 Photoanode

The photoanode exhibits duel functions in PFC system. It applies as photocatalyst and electrocatalyst. It controls the amount of substrate to be decomposed and the production of power density. Therefore, modified materials are used for the preparation of photoanode to improve the efficacy of PFC. Typically, photoanodes applied in PFC are TiO_2, ZnO, CdS/TiO_2, WO_3, AgI/BiOI, α-Fe_2O_3, Fe_2O_3/NiO, $BaTiO_3$, $BiVO_4$, etc. [2, 4, 6–9, 19, 24]. However, TiO_2-based photoanode is mostly employed in PFC. Application of TiO_2 as a photoanode is increased due to its stable physico-chemical characteristics, eco-friendliness, non-toxicity, low price, simple fabrication by chemical or physical techniques, and optoelectronic nature [31]. Titanium oxide can be applied in different forms including nanopowder, nanofilm, nanorods, and nanotubes [57, 58]. The efficiency of nanoscale photoanodes is extremely dependent on their morphology and surface property. The efficiency of photoexcited charge transfer can be increased due to higher surface area of the nanoporus material. Nonetheless, due to the limitation of photocatalytic activity of pure TiO_2 under UV illumination instead of utilizing visible-light, different approaches including metal/metal oxide doping, surface alteration, non-metal doping, dye sensitization, heterojunction construction, and ion implantation are developed to increase the light harvesting capacity in photoanode [28, 59–64].

The modified photoanode materials can enhance the reaction rate and keep the semiconductor corrosion free. Górska *et al.* (2008) [65] reported that various photocatalysts made of TiO_2 were applied in a wide range of temperatures from 350°C to 750°C. They also informed the degradation of phenol using both visible and UV light [65]. Photocatalytic efficiency and absorption of irradiation were also correlated in this study [65]. The highest photocatalytic performance was obtained at 450°C under UV radiation [65]. Monfort *et al.* (2017) [3] described a photocatalysts made of $BiVO_4/TiO_2$ using chemical methods. They examined different sheet formation by $BiVO_4/TiO_2$ coating and observed that the supreme photocatalyst was manufactured by upper-coated TiO_2 sheet on $BiVO_4$ film [3]. This type of composite showed higher photodegradable performance of organic compounds [3]. Sui *et al.* (2017) [53] established a PFC called bio-PFC using TiO_2 nanotubes and Ag nanoparticles, which showed greater degradation of RhB (99.5%) with higher current density (1.26 mA m^{-2}) and power density (318.19 mW m^{-2}) [53]. Tong *et al.* (2017) [10] informed about ZnO-BiOI film synthesized by calcination method that had strong photocatalytic activity under visible light.

He *et al.* (2018) [66] designed a PFC made of TiO_2 and CdS-ZnS that worked under visible light, and higher performance was obtained with increase in electrolyte concentration and light intensity. Based on several studies, it can be stated that the efficacy of photoanode is influenced by gathering of light energy, migration, and separation of photoexcited ion.

5.3.5.2 Photocathode

Photocathode is another major part of PFC. It is constructed with p-type semiconductor. Generally, commercial Pt or Pt-based materials such as carbon paper embedded with platinum-black or carbon-black, Pt wire, Pt foil, etc., are applied as cathode [4, 7, 8, 67]. Although, there are few drawbacks of using Pt. It is costly and rare that leads to major challenges in large-scale application. Pt also shows lesser light harvesting ability and lower photoexcited charge transfer rate. Several investigations have been focused on the development of modified photocathode to enhance the performance of PFC and to make the system more economical.

Recently, various types of cathodes have been developed such as Cu_2O/Cu, ZnO/CuO, homogeneous Cu_2O p-n junction photocathodes, and hematite (α-Fe_2O_3) to reduce the operational cost and to increase the pollutants degradation efficiency [14, 36, 37, 68]. Formation of heterojunction and deposition of co-catalyst are the two efficient approaches to stimulate the chemical reaction, e.g., hydroxide, nickel (II), NiFeOx layers, and cobalt phosphate [69–72]. Graphene oxide (RGO) was applied as an advanced material to associate with semiconductors due to its sufficient active sites. Raptis *et al.* (2018) [52] developed cobalt/nitrogen embedded reduced graphene oxide at 500°C using pyrolysis method that can be applied in the manufacturing of platinum free device. Wang *et al.* (2020) [29] designed a PFC where BiOCl was used as photocathode and furfural was applied as fuel.

5.3.5.3 *Wastewater Treatment and Pollutant Degradation*

In the last few decades, a massive amount of wastewaters have been discharged into the ecosystem that leads to adverse effects on human health and environment. Main constituents of wastewater are heavy metal pollutants, nutrients such as nitrogen and phosphorus, inorganic and organic compounds, dye, etc. Organic compounds or pollutants present in wastewater are the key source of chemical energy that can be converted to electrical energy. As per literature, PFC system has confirmed its efficacy to the treatment of organic pollutants and converted it into nontoxic materials [2, 4–9, 19]. Tang *et al.* (2019) [27] developed a cost-effective PFC using peroxymonosulfate (PMS) and peroxydisulfate (PDS) for the decomposition of Reactive Brilliant Blue (KN-R) as well as power production. Wang *et al.* (2020) [29] designed a PFC where BiOCl was used as photocathode and furfural was applied as fuel. Nordin *et al.* (2020) [25] developed a PFC with joint application of AOP for the removal of organic pollutant and power production under UV light. Occurrence of iron plate had an impact on power generation. Hu *et al.* (2020) [19] fabricated a PFC to remove organic compound and to produce electricity that worked under visible-light that was made of rGO/BiO1-xI (photoanode) and BNC cathode.

The presence of both hydrogen peroxide and Fe^{2+} or Fe^{3+} ion generates more hydroxyl radicals and enhances removal efficiency of organic compounds under UV radiation [13]. This process is called as photo-Fenton processes [13]. To enhance degradation performance, reaction rate, and power generation, few studies have investigated PFC system joint with Fenton process. Zhao *et al.* (2017) [72] designed a PFC termed as Fenton-PFC that was made of TiO_2 as photoanode and ferrous ions were also supplied. Xu *et al.* (2019) [73] informed about a PFC combined with reverse electrodialysis for treatment of wastewater discharging after coal gasification. Higher power generation and H_2O_2 production were related with reverse electrodialysis (RED) stack. Table 5.2 summarizes various types of PFC applied for wastewater treatment and pollutant degradation.

Table 5.2 Various types of PFC used for wastewater treatment and pollutant degradation.

Electrodes (Anode–Cathode)	Organic pollutants	Light source	Electrolyte	Decomposition efficiency (%)	Characteristics	References
ZnO/C–carbon plate	p-Nitrosodimethylaniline (RNO)	UV illumination	0.5 M NaOH	93.4	$P_{max} = 3.14$ mW cm^{-2} at 30-cm^{2} iron plate size	[25]
TiO2/Ti–Pt	Reactive Brilliant Blue (KN-R)	UV illumination	5.0 mmol/L, peroxymonosulfate (PMS) and peroxydisulfate (PDS)	94.2	KN-R removal efficiency was quicker than other process however slower in the power generation.	[27]
BiOCl photocathode–Pt layer as counter electrode Ag/AgCl electrode as reference electrode	Furfural, 5-methyl furfural, and 2-acetylfuran	UV illumination	0.02 mol/L Na_2SO_4	60.2, 43.5, and 43.1	Higher VOC exhibited for Pt-BiOCl furfural fuel cell.	[29]

(Continued)

Table 5.2 Various types of PFC used for wastewater treatment and pollutant degradation. *(Continued)*

Electrodes (Anode–Cathode)	Organic pollutants	Light source	Electrolyte	Decomposition efficiency (%)	Characteristics	References
TiO2/WO3/W (anode)–platinum/BJS (cathode)	Rhodamine B, Methylene blue, Phenol, Bisphenol A, Dichlorophen, and Tetracycline HCl	Sunlight illumination	0.1 M K_2SO_4	99.3, 98.6, 91.7, 95.2, 96.5, and 92.3	Higher performance observed for TiO_2-based PFC than WO_3 photoanode based PFC.	[30]
Ag-TiO_2–Pt	Phenol, tetracycline (TC) 4-chlorophenol (4-CP)	Visible-light	50 mmol/L anhydrous Na_2SO_4	32.6, 37.9, and 96.4	Greater photoactivity observed for photoanode synthesized by powder spray-coating technique at loading proportion 0.62 g/g	[18]

(Continued)

Table 5.2 Various types of PFC used for wastewater treatment and pollutant degradation. (*Continued*)

Electrodes (Anode–Cathode)	Organic pollutants	Light source	Electrolyte	Decomposition efficiency (%)	Characteristics	References
Ag NPs/ ZnO NRs/ NiF–CNFs/ $CoFe_2O_4$/CC	Berberine	Visible light	0.2 mol/L Na_2SO_4 and 0.5 mmol/L PMS solution	93.5	PFC assembled with peroxymonosulfate (PMS) activation process exhibited greater antibiotic removal efficiency with 0.50 V at 2,000 Ω	[71]
(WO_3/ZnO/ Zn)–Pt	Sunset yellow (SSY) and phenol	Sunlight irradiation	1 M Na_2SO_4	-	P_{max} = 0.1052 μW cm^{-2} within 240 min. COD = 84.6% for SSY and 67.5% for phenol	[74]
rGO/BiO_{1-x}Ip–biomass-derived N-embedded carbon (BNC)	Formic acid	Visible light	0.05 M Na_2SO_4 and 0.1 mM $FeSO_4$	68	J_{sc} = 133.18 μA cm^{-2} P_{max} = 17.50 μW cm^{-2}	[19]

(*Continued*)

Table 5.2 Various types of PFC used for wastewater treatment and pollutant degradation. (*Continued*)

Electrodes (Anode–Cathode)	Organic pollutants	Light source	Electrolyte	Decomposition efficiency (%)	Characteristics	References
ZnO/BaTiO3/C/ BaTiO3/ZnO–platinum-loaded carbon	Reactive Red 120 (RR120)	UVA lamps	Electrolyte-free	93.67	P_{max} = 0.5284 mW cm^{-2} at 1:1 of ZnO/ $BaTiO_3$	[26]
TiO_2/ZnO/ Zn-(anode)–TiO_2/CuO/Cu (cathode)	Effluent coming from palm oil mill	UV light irradiation	Sodium sulfate, sodium chloride, and potassium hydroxide	90	V_{oc} = 1173 mV; J_{sc} = 0.2652 mA cm^{-2}; P_{max} = 0.0734 mW cm^{-2}	[20]
TiO_2/Ti–platinized Ti	Reactive Brilliant Red X-3B (RBR) two types of real wastewater collected from textile industry	Mercury lamp	0.05 mol/L Na_2SO_4	90, 62.4, and 50.0	J_{SC} = 0.050 and 0.041 mA cm^{-2}; P_{max} = 0.0113 and 0.0086 mW cm^{-2} for Tex-1 and Tex-2, respectively.	[21]
Fe/GTiP–$ZnIn_2S_4$	Rhodamine B (RhB)	Visible light	0.1 M Na_2SO_4	90	Removal efficiency = 90% at pH 1 with RhB of 10 mg/l	[75]
Burr-like Ag-TiO_2–Pt/C	Brewery effluent	AM1.5 G light source	1 M H_2SO_4 solution with 1 M methanol	COD removal of 14.8%	P_{max} = 1.85 Wm^{-2}	[76]

Table 5.3 Different types of PFC used for electricity generation.

Electrodes (Anode–Cathode)	Organic pollutants	Light source	Electrolyte	Efficiency	References
TiO2 nanotube arrays (TNA)–Graphite	Methylene blue (MB) and Cr(VI)	300-W Xe solar simulator	0.5 M sodium sulfate and sulfuric acid	V_{OC} = 1.06 V; P_{max} = 1 W m^{-2}; J_{SC} = 3.7 A m^{-2}; 96.8% removal efficiency; 95.1% cathodic efficiency within 6 h	[1]
MoS_2/TiO_2/ FTO–MoS_2/Ni	Rhodamine B	Solar light irradiation	0.5 mol/L phosphate-buffered solution (PBS) with 1 mol/L of Na_2SO_3	V_{oc} = 0.7 V; J_{sc} = 1.04 mA cm^{-2}; P_{max} = 0.114 mW cm^{-2}; Degradation efficiency = 69.16% of Rhodamine B at 20 ppm, pH 7	[23]
NiFe-layered double oxide(LDO)/ TiO_2–carbon black	Methanol	Solar light	0.5M Na_2SO_4	V_{oc} = 0.78V; J_{sc} = 1093 μA/cm^{-2}; P_{max} = 169μW/cm^{-2}; Hydrogen peroxide concentration = 0.443 mM after 180 min	[28]

(Continued)

Table 5.3 Different types of PFC used for electricity generation. (*Continued*)

Electrodes (Anode–Cathode)	Organic pollutants	Light source	Electrolyte	Efficiency	References
CdS/ZnS/TiO_2–Pt-black	Ethanol	Visible light	KOH	Current density and power density increased from 5.21 mA cm^{-2} to 11.2 mA cm^{-2} and 1.4 mW cm^{-2} to 4.26 mW cm^{-2}, respectively, with increasing KOH concentration from 0.02 M to 0.5 M	[22]
STNA–Pt-black/Pt	Acetic acid	UV irradiation	0.1 mol/L Na_2SO_4	J_{sc} = 1.42 mA cm^{-2}; V_{oc} = 1.48 V; P_{max} = 0.67 mW cm^{-2}	[24]

(*Continued*)

Table 5.3 Different types of PFC used for electricity generation. (*Continued*)

Electrodes (Anode–Cathode)	Organic pollutants	Light source	Electrolyte	Efficiency	References
g-C_3N_4/WO_3/TiO/Ti–Cu and Cu_2O/Cu	Rhodamine B	Visible light	0.1 M Na_2SO_4	Maximum photocurrent density = 2.27 mA cm^{-2} at 1.1 V; P_{max} = 19.35 μW cm^{-2}; J_{sc} = 0.20 mA cm^{-2}; Degradation efficiency = 87.7%	[78]
TiO2 NRs/FTO–C/Cu NWAs/Cu	Phenol	Simulated solar light	0.1 mol/L Na_2SO_4	Photocurrent = 0.4 mA cm^{-2}; J_{SC} = 0.50 mA cm^{-2}; Removal rate of phenol = 84.2% in 8 h; Hydrogen production = 86.8 μmol cm^{-2}	[55]

(*Continued*)

Table 5.3 Different types of PFC used for electricity generation. (*Continued*)

Electrodes (Anode–Cathode)	**Organic pollutants**	**Light source**	**Electrolyte**	**Efficiency**	**References**
ZnO/CF–Pt/C	Azo dye and Reactive Black 5 (RB5)	Solar light	0.5 M of H_2SO_4	RB5 removal efficiency = 13.6%; V_{oc} = 450 mV; J_{sc} = 0.00058 mA cm^{-2}; P_{max} = 0.1003 mW cm^{-2}	[79]
ZnO/Zn–FeVO4/CF	-	Ultraviolet lamp	0.05 mol/L NaCl	P_{max} = 76 W/m^2; Energy recovery = 16%; Cathodic efficiency = 51%	[73]
TiO_2 NTs/Ti-PPy/Ti	Real textile and wastewater (glucose as fuel)	UV source	0.05 mol/L Na_2SO_4	V_{OC} = 0.46 V; J_{SC} = 1.8–2.2 mA; COD removal rate = 0.0055 min^{-1}	[80]

(*Continued*)

Table 5.3 Different types of PFC used for electricity generation. (*Continued*)

Electrodes (Anode–Cathode)	Organic pollutants	Light source	Electrolyte	Efficiency	References
CdS-ZnS-TiO_2–Pt black	Ethanol	Solar illumination	KOH	V_{OC} increased from 1.06 V to 1.07 V; J_{SC} increased from 1.5 mA/cm^2 to 2.1 mA/cm^2; P_{max} increased from 0.75 mW/cm^2 to 1.01 mW/cm^2 when ethanol concentration increased from 1% to 10%	[66]
WO_3/W–Fe@ Fe_2O_3/carbon felt	MB solution	Xenon lamp (λ = 420 nm)	0.05 mol/L Na_2SO_4	J_{SC} = 0.59 mA/cm^2; P_{max} = 0.34 mW/cm^2 at 20 mg/L of methyl blue	[81]

5.3.6 Electricity Generation

Under light illumination, holes and electrons are formed via photoexcitation process onto photoanode carrying photocatalyst [17, 31]. Photoexcited electrons migrate toward cathode, associate with the holes in the valence band and the reduction of hydrogen ions take place [17, 31]. It creates the alteration in the Fermi level between two electrodes [17]. The efficiency of electricity generation is affected by several features including electrode materials, open-circuit voltage (Voc), short-circuit current (Jsc), fill factor (FF), light intensity, pH of initial solution, nature and concentration of electrolyte, and molecular structure of organic compound [5, 16, 17, 31]. The greater value of FF is proportional to higher efficiency of PFC [17]. Wang *et al.* (2014) [15] designed a solar photocatalytic fuel cell with CdS-TiO_2 photoanode and air-breathing cathode for wastewater treatment as well as electricity generation. A total 10 photodegradable substances, including methanol, formaldehyde propanoic acid, butyric acid, phenol, biphenol A, glycerol, glucose, and xylose, were tested in the PFC reactor [15]. They observed that a short-circuit current of 1.79 mA/cm^2 and a power density of 1134 mW/cm^2 were produced at pH of 4.6 [15]. They also reported a highest short-circuit current of that 5.1 mA/cm^2 and a maximum power density of 3980 mW/cm^2 were attained [15]. Lee *et al.* (2018) [39] assembled a PFC using zinc oxide-zinc as photoanode and platinum-carbon as cathode that worked under sunlight and UV irradiation for the removal of acid orange 7, methylene blue, and Reactive green 19. They found a short-circuit current of 0.0129 mA cm^{-2} and a power density of 0.0032 mW cm^{-2}. Tan *et al.* (2019) [77] developed a PFC using sulfate radicals to reduce organic pollutants in wastewater as well as electricity generation. Short-circuit current density and open-circuit voltage of 3.72 mA cm^{-2} and 1.32 V, respectively, were obtained when the organic pollutant was Methyl orange [77]. Table 5.3 summarizes different types of PFC applied for electricity generation.

5.4 Conclusion

Photocatalytic fuel cell is the combination of photocatalysis process and fuel cell that can be employed for wastewater treatment as well as electricity generation. It can transform chemical energy present in organic pollutants (considered as fuel) into renewable energy via electrochemical process. Performance of PFC system is influenced by several factors such as fabrication and operational technique, types of electrodes, pH of initial solution, nature and concentration of electrolyte, and molecular and chemical structure of organic pollutants. However, PFCs have some issues such as requirement of higher voltage and cost. Maximum research works have been focused on synthesis of modified electrodes for better efficacy of PFC system. However, additional research works will be required to enhance the reusability, stability, performance, and cost effectiveness of PFC system.

Declaration of Interest

The authors declare that they have no known competing financial interests or personal relationships that could have appeared to influence this chapter.

References

1. Liu, X., Xing, Z., Chen, Q., Wang, Y., Multi-functional photocatalytic fuel cell for simultaneous removal of organic pollutant and chromium (VI) accompanied with electricity production. *Chemosphere*, 237, 124457, 2019.
2. Chen, X., Yao, J., Xia, B., Gan, J., Gao, N., Zhang, Z., Influence of pH and DO on the ofloxacin degradation in water by UVA-LED/TiO2 nanotube arrays photocatalytic fuel cell: Mechanism, ROSs contribution and power generation. *J. Hazard. Mater.*, 383, 121220, 2020.
3. Monfort, O., Roch, T., Gregor, M., Satrapinskyy, L., Raptis, D., Lianos, P., Plesch, G., Photooxidative properties of various $BiVO_4/TiO_2$ layered composite films and study of their photocatalytic mechanism in pollutant degradation. *J. Environ. Chem. Eng.*, 5, 5143–5149, 2017.
4. Khalik, W., Ong, S., Ho, L., Wong, Y., Voon, C., Yusuf, S., Yusoff, N., Lee, S., Influence of supporting electrolyte in electricity generation and degradation of organic pollutants in photocatalytic fuel cell. *Environ. Sci. Pollut. Res.*, 23, 16716–16721, 2016.
5. Liu, Y., Li, J., Zhou, B., Li, X., Chen, H., Chen, Q., Wang, Z., Li, L., Wang, J., Cai, W., A TiO_2-nanotube-array-based photocatalytic fuel cell using refractory organic compounds as substrates for electricity generation. *Chem. Commun.*, 47, 10314–10316, 2011.
6. Liu, Y., Li, J., Zhou, B., Lv, S., Li, X., Chen, H., Chen, Q., Cai, W., Photoelectrocatalytic degradation of refractory organic compounds enhanced by a photocatalytic fuel cell. *Appl. Catal. B: Environ.*, 111-112, 485–491, 2012.

7. Sui, M., Dong, Y., You, H., Enhanced photocatalytic activity for the degradation of rhodamine B by integrating salinity gradient power into a photocatalytic fuel cell. *RSC Adv.*, 5, 94184, 2015.
8. Xie, S., Ouyang, K., Ye, X., A novel visible-light responsive photocatalytic fuel cell with a heterostructured $BiVO_4/WO_3$ photoanode and a Pt/C air-breathing cathode. *J. Colloid Interface Sci.*, 532, 758–766, 2018.
9. Zhang, D., Wang, Y., Wang, Y., Zhang, Y., Song, X.M., Fe_2O_3/NiO photocathode for photocatalytic methanol fuel cell: An insight on solar energy conversion. *J. Alloys Compd.*, 815, 152377, 2020.
10. Tong, Y., Zheng, C., Lang, W., Wu, F., Wu, T., Luo, W., Chen, H., ZnO-embedded BiOI hybrid nanoflakes: Synthesis, characterization, and improved photocatalytic properties. *Mater. Des.*, 122, 90–101, 2017.
11. Vasseghian, Y., Khataee, A., Dragoi, N., Moradi, M., Nabavifard, S., Conti, G.O., Khaneghah, A.M., Pollutants degradation and power generation by photocatalytic fuel cells: A comprehensive review. *Arabian J. Chem.*, 13, 8458–8480, 2020.
12. Li, M., Liu, Y., Dong, L., Shen, C., Li, F., Huang, M., Ma, C., Yang, B., An, X., Sand, W., Recent advances on photocatalytic fuel cell for environmental applications-The marriage of photocatalysis and fuel cells. *Sci. Total Environ.*, 668, 966–978, 2019.
13. Singh, M.P. and Pal, H., Concise overview of photocatalytic fuel cell technology for the generation of electricity and degradation of organic wastewater. *Int. Res. J. Eng. Technol.*, 8, 62–67, 2021.
14. Wang, Z., Fan, Y., Wu, R., Huo, Y., Wu, H., Wang, F., Xu, X., Novel magnetic $g\text{-}C_3N_4/a\text{-}Fe_2O_3/Fe_3O_4$ composite for the very effective visible-light-Fenton degradation of Orange II. *RSC Adv.*, 8, 5180, 2018.
15. Wang, B., Zhang, H., Lu, X.Y., Xuan, J., Leung, M.K.H., Solar photocatalytic fuel cell using CdS–TiO_2 photoanode and air-breathing cathode for wastewater treatment and simultaneous electricity production. *Chem. Eng. J.*, 253, 174–182, 2014.
16. Li, J., Li, J., Chen, Q., Bai, J., Zhou, B., Converting hazardous organics into clean energy using a solar responsive dual photoelectrode photocatalytic fuel cell. *J. Hazard. Mater.*, 262, 304–310, 2013.
17. He, Y., Chen, K., Leung, M.K.H., Zhang, Y., Li, L., Li, G., Xuan, J., Li, J., Photocatalytic fuel cell – A review. *Chem. Eng. J.*, 428, 131074, 2022.
18. Deng, B., Fu, S., Zhang, Y., Wang, Y., Ma, D., Dong, S., Simultaneous pollutant degradation and power generation in visible-light responsive photocatalytic fuel cell with an Ag-TiO_2 loaded photoanode. *Nano-Struct. Nano-Objects*, 15, 167–172, 2018.
19. Hu, L., Liao, Y., Xia, D., Peng, F., Tan, L., Hu, S., Zheng, C., Lu, X., He, C., Shu, D., Engineered photocatalytic fuel cell with oxygen vacancies-rich rGO/$BiO_{1-x}I$ as photoanode and biomass-derived N-doped carbon as cathode: Promotion of reactive oxygen species production via Fe^{2+}/Fe^{3+} redox. *Chem. Eng. J.*, 385, 123824, 2020.

20. Kee, M.W., Lam, S.M., Sin, J.C., Zeng, H., Mohamed, A.R., Explicating charge transfer dynamics in anodic TiO_2/ZnO/Zn photocatalytic fuel cell for ameliorated palm oil mill effluent treatment and synchronized energy generation. *J. Photochem. Photobiol. A*, 391, 112353, 2020.
21. Li, K., Zhang, H., Tang, T., Xu, Y., Ying, D., Wang, Y., Jia, J., Optimization and application of TiO_2/Ti-Pt photo fuel cell (PFC) to effectively generate electricity and degrade organic pollutants simultaneously. *Water Res.*, 62, 1–10, 2014.
22. Liao, Q., Li, L., Chen, R., Zhu, X., Wang, H., Ye, D., Cheng, X., Zhang, M., Zhou, Y., Respective electrode potential characteristics of photocatalytic fuel cell with visible-light responsive photoanode and air-breathing cathode. *Int. J. Hydrogen Energy*, 40, 16547–16555, 2015.
23. Li, B., He, Y., Xiao, M., Zhang, Y., Wang, Z., Qin, Z., Chai, B., Yan, J., Li, J., Li, J., Cao, Z., A solar-light driven photocatalytic fuel cell for efficient electricity generation and organic wastewater degradation. *Colloids Surf. A*, 642, 128205, 2022.
24. Liu, Y., Li, J., Zhou, B., Li, X., Chen, H., Chen, Q., Wang, Z., Li, L., Wang, J., Cai, W., Efficient electricity production and simultaneously wastewater treatment via a high-performance photocatalytic fuel cell. *Water Res.*, 45, 3991–3998, 2011.
25. Nordin, N., Ho, L.N., Ong, S.A., Ibrahim, A.H., Rani, A.L.A., Lee, S.L., Ong, Y.P., Hydroxyl radical formation in the hybrid system of photocatalytic fuel cell and peroxi-coagulation process affected by iron plate and UV light. *Chemosphere*, 244, 125459, 2020.
26. Ong, Y.P., Ho, L.N., Ong, S.A., Banjuraizah, J., Ibrahim, A.H., Lee, S.L., Nordin, N., A synergistic heterostructured ZnO/$BaTiO_3$ loaded carbon photoanode in photocatalytic fuel cell for degradation of Reactive Red 120 and electricity generation. *Chemosphere*, 219, 277–285, 2019.
27. Tang, S., Li, N., Yuan, D., Tang, J., Li, X., Zhang, C., Rao, Y., Comparative study of persulfate oxidants promoted photocatalytic fuel cell performance: Simultaneous dye removal and electricity generation. *Chemosphere*, 234, 658–667, 2019.
28. Tao, S., Wang, F., Zhang, J., Shi, J., Guo, W., Lu, J., Visible Light Responsive TiO_2/NiFe Mixed Metal Oxide Carbon Photocatalytic Fuel Cell with Synchronous Hydrogen Peroxide Production. *Eur. J. Inorg. Chem.*, 13, 1230–1239, 2021.
29. Wang, Y., Wang, Y., Song, X., Zhang, Y., Ma, T., BiOCl-based photocathode for photocatalytic fuel cell. *Appl. Surf. Sci.*, 506, 144949, 2020.
30. Zeng, Q., Bai, J., Lia, J., Li, L., Xia, L., Zhou, B., Sun, Y., Highly-stable and efficient photocatalytic fuel cell based on an epitaxial TiO_2/WO_3/W nanothorn photoanode and enhanced radical reactions for simultaneous electricity production and wastewater treatment. *Appl. Energy*, 220, 127–137, 2018.

31. Oli, H.B., Kim, A., Park, M., Bhattarai, D.P., Pant, B., Photocatalytic Fuel Cells for Simultaneous Wastewater Treatment and Power Generation: Mechanisms, Challenges, and Future Prospects. *Energies*, 15, 3216, 2022.
32. Zhang, J., Tian, B., Wang, L., Xing, M., Lei, J., Mechanism of Photocatalysis, in: *Photocatalysis*, pp. 1–15, 2018.
33. Saravanan, R., Gracia, F., Stephen, A., Basic Principles, Mechanism, and Challenges of Photocatalysis, in: *Nanocomposites for Visible Light-induced Photocatalysis*, pp. 19–40, 2017.
34. Selihin, N.M. and Tay, M.G., A review on future wastewater treatment technologies: Micro-nanobubbles, hybrid electro-Fenton processes, photocatalytic fuel cells, and microbial fuel cells. *Water Sci. Technol.*, 85, 319, 2022.
35. Ge, J., Zhang, Y., Heo, Y.J., Park, S.J., Advanced Design and Synthesis of Composite Photocatalysts for the Remediation of Wastewater: A Review. *Catalysts*, 9, 122, 2019.
36. Bai, J., Wang, R., Li, Y., Tang, Y., Zeng, Q., Xia, L., Li, X., Li, J., Li, C., Zhou, B., A solar light driven dual photoelectrode photocatalytic fuel cell (PFC) for simultaneous wastewater treatment and electricity generation. *J. Hazard. Mater.*, 311, 51–62, 2016.
37. Chen, Q., Li, J., Li, X., Huang, K., Zhou, B., Cai, W., Shangguan, W., Visible-Light Responsive Photocatalytic Fuel Cell Based on WO_3/W Photoanode and Cu_2O/Cu Photocathode for Simultaneous Wastewater Treatment and Electricity Generation. *Environ. Sci. Technol.*, 46, 11451–11458, 2012.
38. Pan, H., Liao, W., Sun, N., Murugananthan, M., Zhang, Y., Highly Efficient and Visible Light Responsive Heterojunction Composites as Dual Photoelectrodes for Photocatalytic Fuel Cell. *Catalysts*, 8, 30, 2018.
39. Lee, S.L., Ho, L.N., Ong, S.A., Wong, Y.S., Voon, C.H., Khalik, W.F., Yusoff, N.A., Nordin, N., Exploring the relationship between molecular structure of dyes and light sources for photodegradation and electricity generation in photocatalytic fuel cell. *Chemosphere*, 209, 935–943, 2018.
40. Fujishima, A. and Honda, K., Electrochemical photolysis of water at a semiconductor electrode. *Nature*, 238, 37–38, 1972.
41. Kaneko, M., Gokan, N., Katakura, N., Takeia, Y., Hoshino, M., Artificial photochemical nitrogen cycle to produce nitrogen and hydrogen from ammonia by platinized TiO_2 and its application to a photofuel cell. *Chem. Commun.*, 1625-1627, 1625, 2005.
42. Lianos, P., Production of electricity and hydrogen by photocatalytic degradation of organic wastes in a photoelectrochemical cell. The concept of the Photofuelcell: A review of a re-emerging research field. *J. Hazard. Mater.*, 185, 575–590, 2011.
43. Li, J., Bai, J., Niu, X., Li, X., Chen, S., Wang, J., Zhou, B., Dramatic enhancement of photocurrent for $BiVO_4$/TiO_2 heterojunction photoanode with suitable band-match via in-situ band regulation using Ta. *Int. J. Hydrogen Energy*, 43, 18202–18210, 2018.

44. Li, N., Tang, S., Rao, Y., Qi, J., Wang, P., Jiang, Y., Huang, H., Gu, J., Yuan, D., Improved dye removal and simultaneous electricity production in a photocatalytic fuel cell coupling with persulfate activation. *Electrochim. Acta*, 270, 330–338, 2018.
45. Li, X., Liu, S., Cao, D., Mao, R., Zhao, X., Synergetic activation of H_2O_2 by photogenerated electrons and cathodic Fenton reaction for enhanced self-driven photoelectrocatalytic degradation of organic pollutants. *Appl. Catal. B: Environ.*, 235, 1–8, 2018.
46. Liu, Y., Zheng, Y., Du, B., Nasaruddin, R.R., Chen, T., Xie, J., Golden carbon nanotube membrane for continuous flow catalysis. *Ind. Eng. Chem. Res.*, 56, 2999–3007, 2017.
47. Fu, J., Yu, J., Jiang, C., Cheng, B., g-C3N4-based heterostructured photocatalysts. *Adv. Energy Mater.*, 8, 1701503, 2018.
48. Zhou, Z., Wei, L., Wang, Y., Karahan, E., Chen, Z., Lei, Y., Chen, X., Zhai, S., Liaod, X., Chen, Y., Hydrogen evolution reaction activity of nickel phosphide is highly sensitive to electrolyte pH. *J. Mater. Chem. A*, 5, 20390, 2017.
49. Shi, X., Back, S., Gill, T.M., Siahrostami, S., Zheng, X., Electrochemical Synthesis of H_2O_2 by Two-Electron Water Oxidation Reaction. *Chem.*, 7, 38–63, 2021.
50. Ohtani, B., Photocatalysis by inorganic solid materials: Revisiting its definition, concepts, and experimental procedures. *Adv. Inorg. Chem.*, 63, 395–430, 2011.
51. Balis, N., Dracopoulos, V., Antoniadou, M., Lianos, P., One-step electrodeposition of polypyrrole applied as oxygen reduction electrocatalyst in Photoactivated Fuel Cells. *Electrochim. Acta*, 70, 338–343, 2012.
52. Raptis, D., Ploumistos, A., Zagoraiou, E., Thomou, E., Daletou, M., Sygellou, L., Tasis, D., Lianos, P., Co-N doped reduced graphene oxide as oxygen reduction electrocatalyst applied to Photocatalytic Fuel Cells. *Catal. Today*, 315, 31–35, 2018.
53. Sui, M., Dong, Y., Wang, Z., Wang, F., You, H., A biocathode-driven photocatalytic fuel cell using an Ag-doped TiO_2/Ti mesh photoanode for electricity generation and pollutant degradation. *J. Photochem. Photobiol. A*, 348, 238–245, 2017.
54. Xia, L., Bai, J., Li, J., Zeng, Q., Li, X., Zhou, B., A highly efficient $BiVO_4$/WO_3/W heterojunction photoanode for visible-light responsive dual photoelectrode photocatalytic fuel cell. *Appl. Catal. B: Environ.*, 183, 224–230, 2016.
55. Wu, Z., Zhao, G., Zhang, Y., Liu, J., Zhang, Y., Shi, H., A solar-driven photocatalytic fuel cell with dual photoelectrode for simultaneous wastewater treatment and hydrogen production. *J. Mater. Chem. A*, 3, 3416–3424, 2015.
56. Rabé, K., Lifen Liu, L., Nahyoon, N.A., Zhang, Y., Idris, A.M., Visible-light photocatalytic fuel cell with Z-scheme g-C_3N_4/Fe^0/TiO_2 anode and WO_3 cathode efficiently degrades berberine chloride and stably generates electricity. *Sep. Purif. Technol.*, 212, 774–782, 2019.

57. Bhattarai, D.P., Shrestha, S., Shrestha, B.K., Park, C.H., Kim, C.S., A controlled surface geometry of polyaniline doped titania nanotubes biointerface for accelerating MC3T3-E1 cells growth in bone tissue engineering. *Chem. Eng. J.*, 350, 57–68, 2018.
58. Leong, S., Razmjou, A., Wang, K., Hapgood, K., Zhang, X., Wang, H., TiO_2 based photocatalytic membranes: A review. *J. Membr. Sci.*, 472, 167–184, 2014.
59. Dubey, P.K., Kumar, R., Tiwari, R.S., Srivastava, O.N., Pandey, A.C., Singh, P., Surface modification of aligned TiO_2 nanotubes by Cu_2O nanoparticles and their enhanced photo electrochemical properties and hydrogen generation application. *Int. J. Hydrogen Energy*, 43, 6867–6878, 2018.
60. Marami, M.B., Farahmandjou, M., Khoshnevisan, B., Sol–Gel Synthesis of Fe-Doped TiO_2 Nanocrystals. *J. Electron. Mater.*, 47, 3741–3748, 2018.
61. Meng, S., Zhang, J., Chen, S., Zhang, S., Huang, W., Perspective on construction of heterojunction photocatalysts and the complete utilization of photogenerated charge carriers. *Appl. Surf. Sci.*, 476, 982–992, 2019.
62. Qin, L.Z., Liang, H., Liao, B., Liu, A.D., Wu, X.Y., Sun, J., Photocatalytic performance of Fe-, Ni-, or Cu-ion implanted TiO_2 films under UV light, visible light and sunlight irradiation. *Nucl. Instrum. Methods Phys. Res., Sect. B*, 307, 385–390, 2013.
63. Thor, S.H., Ho, L.N., Ong, S.A., Abidin, C.Z.A., Heah, C.Y., Nordin, N., Ong, Y.P., Yap, K.L., Advanced oxidation treatment of amaranth dye synchronized with electricity generation using carbon-based cathodes in a sustainable photocatalytic fuel cell integrated electro-fenton system. *J. Environ. Chem. Eng.*, 9, 106439, 2021.
64. Yu, H., Zhang, M., Wang, Y., Lv, J., Liu, Y., He, G., Sun, Z., Low-temperature strategy for vapor phase hydrothermal synthesis of C\N\S-doped TiO_2 nanorod arrays with enhanced photoelectrochemical and photocatalytic activity. *J. Ind. Eng. Chem.*, 98, 130–139, 2021.
65. Górska, P., Zaleska, A., Kowalska, E., Klimczuk, T., Janusz, Sobczak, J.W., Skwarek, E., Janusz, W., Hupka, J., TiO_2 photoactivity in vis and UV light: The influence of calcination temperature and surface properties. *Appl. Catal. B: Environ.*, 84, 440–447, 2008.
66. He, X., Chen, M., Chen, R., Zhu, X., Liao, Q., Ye, D., Zhang, B., Zhang, W., Yu, Y., A solar responsive photocatalytic fuel cell with the membrane electrode assembly design for simultaneous wastewater treatment and electricity generation. *J. Hazard. Mater.*, 358, 346–354, 2018.
67. Lee, S.L., Ho, L.N., Ong, S.A., Wong, Y.S., Voon, C.H., Khalik, W.F., Yusoff, N.A., Nordin, N., Enhanced electricity generation and degradation of the azo dye Reactive Green 19 in a photocatalytic fuel cell using ZnO/Zn as the photoanode. *J. Cleaner Prod.*, 127, 579–584, 2016.
68. She, X., Wu, J., Xu, H., Zhong, J., Wang, Y., Song, Y., Nie, K., Liu, Y., Yang, Y., Rodrigues, M.F., Vajtai, R., Lou, J., Du, D., Li, H., Ajayan., P.M., High

Efficiency Photocatalytic Water Splitting Using 2D α-Fe_2O_3/g-C_3N_4 Z-Scheme Catalysts Adv. *Energy Mater.*, 7, 1700025, 2017.
69. Young, K.M.H. and Hamann, T.W., Enhanced photocatalytic water oxidation efficiency with $Ni(OH)_2$ catalysts deposited on a-Fe_2O_3 via ALD. *Chem. Commun.*, 50, 8727–8730, 2014.
70. Klahr, B., Gimenez, S., Fabregat-Santiago, F., Bisquert, J., Hamann, T.W., Photoelectrochemical and Impedance Spectroscopic Investigation of Water Oxidation with "Co-Pi"-Coated Hematite Electrodes. *J. Am. Chem. Soc*, 134, 16693–16700, 2012.
71. Zhang, Y., Liu, L., Chen, Q., Yun He, Y., Leung, M.K.H., Electricity generating & high efficiency advanced oxidation process including peroxymonosulfate activation in photocatalytic fuel cell. *Chem. Eng. J.*, 378, 122148, 2019.
72. Zhao, K., Zeng, Q., Bai, J., Li, J., Xia, L., Chen, S., Zhou, B., Enhanced organic pollutants degradation and electricity production simultaneously via strengthening the radicals reaction in a novel Fenton-photocatalytic fuel cell system. *Water Res.*, 108, 293–300, 2017.
73. Xu, P. and Xu, H., Enhanced Electricity Generation and H_2O_2 Production in a Photocatalytic Fuel Cell and Fenton Hybrid System Assisted with Reverse Electrodialysis. *ACS Omega*, 4, 5848–5851, 2019.
74. Lam, S.M., Sin, J.C., Lin, H., Li, H., Lim, J.W., Zeng, H., A Z-scheme WO_3 loaded-hexagonal rod-like ZnO/Zn photocatalytic fuel cell for chemical energy recuperation from food wastewater treatment. *Appl. Surf. Sci.*, 514, 145945, 2020.
75. Nahyoon, N.A., Liu, L., Rabe, K., Thebo, K.H., Yuan, L., Sun, J., Yang, F., Significant photocatalytic degradation and electricity generation in the photocatalytic fuel cell (PFC) using novel anodic nanocomposite of Fe, graphene oxide, and titanium phosphate. *Electrochim. Acta*, 271, 41–48, 2018.
76. Lui, G., Jiang, G., Fowler, M., Yu, A., Chen, Z., A high performance wastewater-fed flow-photocatalytic fuel cell. *J. Power Sources*, 425, 69–75, 2019.
77. Tan, X., Bai, J., Zheng, J., Zhang, Y., Li, J., Zhou, T., Xia, L., Xu, Q., Zhou, B., Photocatalytic fuel cell based on sulfate radicals converted from sulfates in situ for wastewater treatment and chemical energy utilization. *Catal. Today*, 335, 485–491, 2019.
78. Yao, H., Xu, Y., Zhong, D., Zeng, Y., Zhong, N., Efcient rhodamine B degradation and stable electricity generation performance of visible□light photocatalytic fuel cell with g□C_3N_4/WO_3/TiO_2/Ti photoanode. *Ionics*, 27, 4875–4884, 2021.
79. Khalik, W.F., Ho, L., Ong, S.A., Voon, C.H., Wong, Y.S., Yusuf, S.Y., Yusoff, N., Lee, S.L., Reactive Black 5 as electron donor and/or electron acceptor in dual chamber of solar photocatalytic fuel cell. *Chemosphere*, 202, 467–475, 2018.

80. Li, K., Zhang, H., Tang, T., Tang, Y., Wang, Y., Jia, J., Facile electrochemical polymerization of polypyrrole film applied as cathode material in dual rotating disk photo fuel cell. *J. Power Sources*, 324, 368–377, 2016.
81. Xu, P., Xu, H., Zheng, D., Simultaneous electricity generation and wastewater treatment in a photocatalytic fuel cell integrating electro-Fenton process. *J. Power Sources*, 421, 156–161, 2019.

6

Remediation of Wastewater through Photo-Induced Catalytic and Electrochemical Hydrogen Production

Ramgopal Tiwari, Ajayrajsinh R. Zala and Premlata Kumari*

Department of Chemistry, Sardar Vallabhbhai National Institute of Technology, Surat, Gujarat, India

Abstract

Human activity in industrialization has raised both the amount of waste produced and the need for energy in recent years. An appealing method to both, reducing environmental pollution and supply energy, is the treatment of contaminants found in wastewater in conjunction with energy recovery. The exhaustiveness of non-renewable fuels compelled researchers to find a potent and renewable fuel alternative. Sunlight being ubiquitous and renewable energy has grasped many recent research studies and inventions. Production of hydrogen green fuel photocatalytically using photoelectrodes has proved to be a promising approach. Hydrogen serves as a clean energy source for the shift to a decarbonized civilization. Hydrogen can be produced from contaminants that have been hydrogenated in water using photocatalysis; the whole reaction is more favorable thermodynamically than water splitting for hydrogen. This book chapter focuses on the recent advances in the study of hydrogen generation from wastewater contaminants, photocatalytically and photoelectrochemically. In addition, materials, efficiency, and the fundamentals of photocatalysis and photoelectrochemical cells is also covered. Further, the discussion emphasizes on how the oxidation of chemicals found in wastewater leads to hydrogen generation.

Keywords: Wastewater treatment, photocatalysis, photoelectrochemical cell, hydrogen generation

**Corresponding author*: pl@chem.svnit.ac.in

Sadanand Pandey, Elvis Fosso-Kankeu and Soumya Pandit (eds.) Synergy of Bio-Chemical Processes for Photocatalytic and Photoelectrochemical Wastewater Treatment, (139–160) © 2024 Scrivener Publishing LLC

List of Abbreviations

PET	photoelectron transfer
AOP	advanced oxidation process
EDTA	hexamethylene tetra acetic acid
OA	oxalic acid
NTA	nitrilotriacetic acid
CA	carbonic acid
EDDS	Ethylenediamine-N,N′-disuccinic acid
COD	chemical oxygen demand
BOD	biological oxygen demand
QY	quantum yield
MOF	metal–organic framework
POP	porous organic polymers
FAME	fatty acid methyl ester
PCET	proton-coupled electron transfer
PEC	photoelectrochemical cell
AC	activated carbon

6.1 Introduction

Employing innovative and cutting-edge chemical transformation techniques is important due to industries' rapid and continuous evolution. These are particularly crucial because such industrial expansion has resulted in a rise in pollutant generation [1]. Several harmful substances are released into the water resources due to daily human activities, like processing chemicals in industries and agriculture. The globe's population, which is over half, is in regions with poor access to potable water [2]. Consequently, providing clean water to everyone on the planet is a challenging task exacerbated by population growth and climate change [3]. Contaminated water poses a significant hazard to all life on Earth. So, before being discharged into the environment, industrial effluents must be treated for any still present pollutants. The removal of contaminants is currently accomplished using techniques including filtering, coagulation, membrane separation, adsorption, and biological processes [4–8]. Due to their ineffectiveness, high expense, and requirement of complicated apparatus, these procedures are unsuitable [9]. Photocatalytic degradation is an effective wastewater treatment technique that does not produce secondary pollution and is an alternative to traditional water treatment techniques. Hydrogen is a clean energy source that can be produced from contaminants hydrogenated in water using photocatalysis. A variety of photocatalysts, such as metal oxides like ZnO, TiO_2, NiO, and WO_3 [10–12], halides like BiOI and AgCl [13], metal sulfides like CdS, ZnS, In_2S_3, and MoS_2 [14, 15], and non-metal semiconductors like g-C_3N_4 and graphene [16], are used to remove impurities from aqueous solutions.

An incoming photon can be converted into a suitable or storable energy source by a photocatalytic material thanks to its propensity to produce an electron/hole pair at the photocatalyst level.

A photocatalyst, in a broad sense, is a catalyst that, when exposed to light, manifests its catalytic properties through photon absorption. The key phenomenon in photocatalysis is photo-induced electron transfer (PET) resulting from the absorption of photons having energy greater than the energy gap between the valence band and the conduction band. PET may lead to the following processes:

1. **Photovoltaic process:** Electron–hole pairs are produced by photocatalytic materials, especially semiconductors when they absorb enough energy to excite their electrons from the valence band to the conduction band, which can be exploited to produce electricity. For instance, the photovoltaic method is employed in solar cells.
2. **Photocatalysis process:** It is the process in which the generated electron–hole pairs are employed to promote chemical reactions. For instance, utilizing photocatalysts to remediate wastewater.
3. **Super hydrophilicity:** when the surface properties are altered using generated electron–hole pairs. For instance, when light is irradiated onto titanium dioxide, the surface characteristics are changed, creating zero contact angle, defogged glass, etc.

Advanced oxidation processes (AOPs) include the photocatalytic water treatment method, a more prominent family of processes [17]. To fragment and degrade organic molecules, all AOPs entail the generation and use of highly oxidative OH· radicals. The compounds that could be addressed include any residual xenobiotics, dyes, pesticides, or herbicides consumed by humans or used in industrial processes and ended up in wastewater [18]. For instance, photolytic and photo-Fenton techniques are widely used in modern wastewater treatment facilities. By subjecting sacrificial substances like ozone (O_3) or hydrogen peroxide (H_2O_2) to intense germicidal ultraviolet (UV) radiation (254 nm), oxidative radicals are produced in photolytic reactions. In the photo-Fenton process, catalytic ferrous ions (Fe^{2+}), H_2O_2, and UV light are combined to more promptly produce OH· radicals. Due to the requirement of high-energy UV light, this procedure has the disadvantage of being cost-prohibitive.

Heterogeneous photocatalysis is one approach that might be used to overcome these restrictions. Simple photocatalysts in the form of particles or nanoparticles supported on a substrate such as a membrane or suspended in water are utilized in this process instead of sacrificial or additional chemicals.

6.2 Mechanism of Photocatalysis

Photoexcitation of the valence electron to the conduction band, leaving a hole in the valence band, occurs when a semiconductor photocatalyst is exposed to photons with energy equal to or greater than the bandgap between the highest energy occupied band (valence band) and the lowest energy unoccupied band (conduction band). The resultant electron–hole pair has a lifetime measured in nanoseconds and, if the electron cascades down to a metastable state, in microseconds. This equips us with an opportunity to carry out chemical reactions by utilizing the excited electron to react with organic substrates like dyes and other effluents. The recombination of excited electrons with holes in the valence band, however, is another potential outcome and can also result in the input energy being dissipated as heat energy. However, it is possible to prevent the recombination of electron–hole pairs by using the appropriate electron–hole scavengers. Figure 6.1 shows how photo-induced holes can be used to oxidize substrates. Holes acquired at the semiconductor

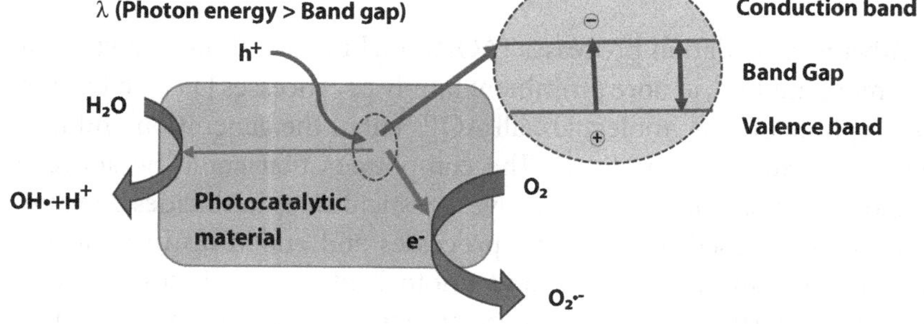

Figure 6.1 Schematic depiction of photocatalytic initiation process [19] (https://doi.org/10.1039/C5TA01637A).

surface react with OH^- to produce OH•, a potent-oxidizing agent capable of effectively degrading organic contaminants like dyes and hydrocarbons into safe final products (e.g., CO_2 and H_2O). On the other hand, the superoxide anion (O_2^-) produced by the reduction of molecular oxygen (O_2) through the photoexcited electron on the photocatalyst surface induces the photodegradation of organic contaminants as effectively as a hole. The semiconductor is unaffected by any of these processes, and there is constant charge transfer to the adsorbate.

Reactions:

$$\text{Photocatalyst} + hv = \text{Photocatalyst } [h_{(vb)}^{+} + e_{(cb)}^{-}]$$

$$H_2O^- + h_{(vb)}^{+} \rightarrow H^+ + OH^{\bullet}$$

$$e_{(cb)} + O_2 \rightarrow O_2^{\bullet-}$$

$$O_2^{\bullet-} + H^+ \rightarrow HO_2^{\bullet}$$

$$O_2^{\bullet-} + HO_2^{\bullet} + H^+ \rightarrow H_2O_2 + O_2$$

$$H_2O_2 + hv \rightarrow 2OH\bullet$$

$$OH^{\bullet} + \text{Organic compounds} \rightarrow CO_2 + H_2O + \text{by-product}$$

6.3 Homogeneous and Heterogeneous Photocatalysis

6.3.1 Homogeneous Photocatalysis

6.3.1.1 Fenton Reaction

Fenton in 1894 discovered the Fenton reaction that described the oxidation of tartaric acid by H_2O_2 and catalyzed by ferrous (Fe^{2+}) salts. The Fenton procedure was extensively expanded by the late 1960s to eliminate harmful organic contaminants in the effluents [20]. The Fenton method was primarily employed for the flocculation and radical oxidation of wastewater. The oxidation of refractory organic pollutants into harmless molecules like CO_2, water, and inorganic salts was catalyzed by ferrous ions to produce HO· (standard potential = 2.80 V versus standard hydrogen electrode), which can also cause the production of other radicals that can completely oxidize organic materials.

$$Fe^{2+} + H_2O_2 \rightarrow Fe^{3+} + \bullet OH + OH^-$$

$$Fe^{2+} + \bullet OH \rightarrow Fe^{3+} + OH^-$$

$$RH + \bullet OH \rightarrow H_2O + R \rightarrow \text{further oxidation}$$

$$R\bullet + Fe^{3+} \rightarrow Fe^{2+} + R^+$$

The conventional Fenton process requires a strict pH range, high H_2O_2 consumption, and produces too much sludge that affects the oxidation efficiency of hydrogen peroxide. This can be abrogated by using reductant or transition metal or by utilizing some energy sources like light (photo-Fenton process) and electricity (electro-Fenton process) to eliminate iron usage and the consequent production of sludge.

6.3.1.1.1 Photo-Fenton Process

In the presence of UV light, the Fenton process evolves to produce more hydroxyl ions. The photo-Fenton reaction involves the photochemical regeneration of Fe^{2+} by the reduction of ferric Fe^{3+}. The freshly produced Fe^{2+} combines with H_2O_2 to produce 2OH free radicals and Fe^{3+}, and the cycle continues as depicted below:

$$Fe^{2+} + H_2O_2 \rightarrow Fe^{3+} + 2OH^\bullet$$

$$Fe(OH)^{2+} + hv \rightarrow Fe^{2+} + {}^\bullet OH$$

$$H_2O_2 + hv \rightarrow 2{}^\bullet OH$$

Chelating agents have improved the oxidation efficiency by complexing Fe^{3+} in the order EDTA > OA > NTA > CA > EDDS, but this also increases the chemical oxygen demand (COD) of the water [21].

Advantages of the photo-Fenton procedure over the classical Fenton process are as follows:

- A photo-induced Fe^{3+}/Fe^{2+} redox cycle could decrease the dosage of the catalyst in the Fenton process, thereby reducing sludge formation.
- Solar radiation increases the utilization of H_2O_2, improving the atom economy and incorporating green chemistry principles.

6.3.1.1.2 Electro-Fenton Process

The electrochemical-Fenton–coupled method produces Fe^{2+} and H_2O_2 (Fenton reagent) at the two electrodes, leading to improved degradation of wastewater pollutants.

The series of reactions at the anode and cathode of the electrochemical cell is listed below:

$$O_2 + 2H+ + 2e- \rightarrow H_2O_2$$

$$Fe^{3+} + e^- \rightarrow Fe^{2+}$$

$$2H_2O \rightarrow O_2 + 4H^+ + 4e^-$$

$$H_2O + H_2O_2 \rightarrow H^- + \rightarrow Fe^{2+} + \bullet OH + e^-$$

The significant advantages of the electro-Fenton process over the conventional method are as follows:

- The production of H_2O_2 *in situ* by electrochemical methods has advantages because it increases the effectiveness of organic pollutant degradation, lowers the cost, and lowers transportation-related dangers.
- The production of iron sludge is lowered as a consequence of the regeneration of ferrous ions (Fe^{2+}) by the reduction of ferric ions on the cathode.

6.3.2 Heterogeneous Photocatalysis

This category of photocatalysis typically consists of semiconductor materials, transition metal oxide, and their doped alloys, with TiO_2 oxide being the most important, studied, and used heterogeneous photocatalytic material for almost completely degrading pollutants into CO_2 and other simpler benign molecules, which can take place at ambient conditions. TiO_2 has a wide range of applications because it is reusable, economical, and non-toxic.

For the degradation of organic contaminants, heterogeneous photocatalysis uses a series of redox processes at the photocatalyst level. The free ·OH radicals generated due to redox reaction decompose complex toxic pollutants into simpler non-toxic gases and compounds. The reactions involved are shown below [22]:

Photocatalytic materials:

$$TiO_2 + hv \rightarrow h^+_{VB} + e^-_{CB}$$

$$e^-_{CB} + h^+_{VB} + \rightarrow \text{energy}$$

$$H_2O + h^+_{VB} \rightarrow \bullet OH + H^+$$

$$O_2 + e^-_{CB} \rightarrow O^-_2$$

6.3.2.1 Photocatalytic Materials

Photocatalytic materials have immense importance not only for their impeccable role in decomposing water and toxic and harmful effluents but also for their ability to generate electricity from insolation received. The widely used photocatalytic materials include n-type semiconductor materials like TiO_2, SnO_2, ZnO [23], Fe_2O_3, WO_3, and In_2O_3 as well as non-metallic semiconductors like C_3N_4 [24]. Among all, titanium dioxide (TiO_2) and zinc oxide (ZnO) [25] stand at the front owing to their low-cost preparation, robustness, high surface area, and non-toxicity.

Titanium dioxide (TiO_2) has been a priority since 1972 as it has a small bandgap equivalent to 3.2 eV. It makes electron–hole pair formation feasible and further reactions favorable, with high photocatalytic efficiency, good optical and chemical stability, and optimal environmental and biological compatibility. ZnO also has comparable bandgap energy to TiO_2; its higher electron mobility and lifetime, relatively lower production costs, and easy fabrication make it a potent photocatalytic material. Table 6.1 illustrates the bandgap of various metal oxides. The world of photocatalysts witnessed a new class of composites, $K_4Nb_6O_{17}$, with layered structured and potent desired properties for photocatalysis, before which the photocatalyst market was largely occupied with metal oxides like $SrTiO_3$, TiO_2, and other tantalates [26]. ZrO_2 is also one of the photocatalysts which exhibit high activity without the assistance of a co-catalyst [27]. In addition, biopolymer-supported photocatalytic metal oxides are used like TiO_2 (most used biopolymers for TiO_2 support are chitosan [28], alginate [29], cellulose [30], cyclodextrin [31], and collagen and starch [32]) are used as photocatalysts for detoxification of environment. Many new nanomaterials are also in development and used as photocatalysts.

Table 6.1 Bandgaps of various semiconducting metal oxides [Water Air Soil Pollut (2023) 234:349, https://doi.org/10.1007/s11270-023-06359-9].

Metal oxides	Bandgap (eV)
ZrO_2	5.0
MnO	3.8
SnO_2	3.7
ZnO	3.4
CeO_2	3.2
FeO	2.4
Fe_2O_3	2.0
MnO_2	1.3
WO_3	2.7
TiO_2	3.0

6.3.2.2 *Applications of Photocatalytic Processes*

- **Water splitting:** Photocatalytic water splitting uses the process of PET for dissociating water molecules into hydrogen (H_2) and oxygen (O_2) using a catalyst sensitive to light as shown in Figure 6.2. The produced hydrogen is a green and renewable fuel; hence, water splitting is of utmost interest to chemists and technologists all over the globe. The process is endothermic ($\Delta H > 0$) in nature because a considerable amount of energy is needed to break the O-H bond present in the water molecules [33].

The photocatalyst provides a surface splitting of the water molecules by absorbing photons (UV and visible light) to undergo photoexcitation and then combining with H^+ to give H_2. The relative activity of photocatalysts can be measured by comparing their quantum yields (QY), which are given as follows:

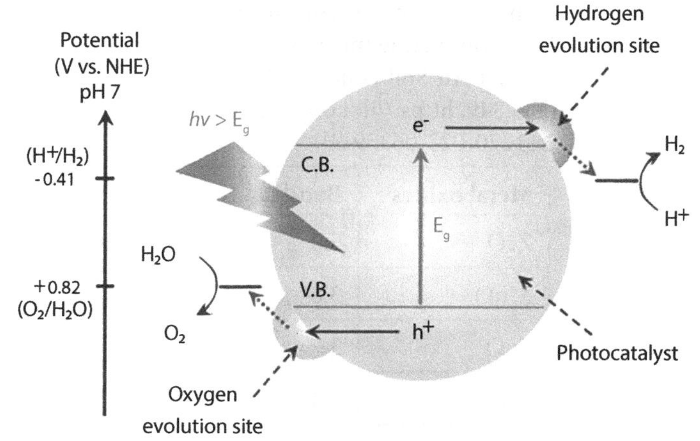

Figure 6.2 Photocatalytic water splitting (https://pubs.acs.org/doi/10.1021/jz1007966).

$$QY\ (\%) = (\text{Photochemical reaction rate}) / (\text{Photon absorption rate}) \times 100\%.$$

A photocatalyst with high QY will give a high rate of H_2 evolution and will be a better catalyst.

The various suitable photocatalysts for water splitting are as follows:

$Cd_{1-x}Zn_xS$: Generally used in producing H_2 from aqueous solutions containing SO_3^{2-}/S^{2-} as a sacrificial agent under the light. The highest activity is seen when the concentration of Zn is increased from 0.2 to 0.3, leading to increased production of H_2 gas [34].

$NaTaO_3$:La: This nano-step structured material has edges assisting in H_2 production and grooves liberating O_2 from H_2O oxidation. It has one of the highest water splitting rates among photocatalysts, with splitting rates of 9.7 mmol/h and QY of 56%. An increase in H_2 production was observed by adding NiO as a co-catalyst [9].

$(Ga_{.82}Zn_{.18})$ $(N_{.82}O_{.18})$ and $K_3Ta_3B_2O_{12}$ are other examples of semiconductor-doped metal photocatalysts that use UV or light to generate H_2 gas. Some different photocatalysts include the following: molecular catalysts: bismuth vanadate [35], bismuth oxides [36], and tungsten diselenide (WSe_2) [37]; and aluminium-based metal–organic frameworks (MOF): porous organic polymers (POPs) [38] and ansa-titanocene(III/IV) triflate complexes [39].

Fly ash usage as a catalyst supports the photodegradation of organic contaminants and reforming processes for the ensuing wastewater remediation and H_2 energy generation [40].

Due to their optical band of 3.25 eV, $Bi_4Ti_3O_{12}$ nanostructured materials produce less hydrogen. The use of dopant materials, such as CdS and PbS compounds, can increase efficiency [41, 42].

Solar cells: Solar cells produce electricity by absorbing photons, which are tiny packets of energy from the sun that are subsequently absorbed by photocatalytic materials. Numerous kinds of solar cells, including photoelectrochemical, organic, color-sensitive, and hybrid cells, have been developed to harness the energy of the sun [43].

- **Reduction of carbon dioxide:** Industrialization and technological advancement have led to increased greenhouse gases that must be abrogated. Photocatalytic reduction of CO_2 into hydrocarbon fuels has recently been a major player. The hydrocarbon, particularly methanol produced during the photocatalytic reduction, is a major feedstock for many secondary industries and can be used in making biodiesel by transesterification reaction as in FAME (fatty acid methyl esters). Photocatalytic reduction of CO_2 is analogous to the process of photosynthesis performed by green plants to convert CO_2 and water into carbohydrates and oxygen when chlorophyll, the natural photocatalyst, absorbs incoming solar energy to convert it into the chemical energy of saccharides. The photocatalytic reduction is generally done on the semiconductor surface, and its mechanism is akin to the water-splitting reaction. The significant steps involved are the following:

    - Absorption of light photons by the photocatalyst
    - Separating charge carriers by photoexcited electrons from VB to CB leads to the forming of electron–hole pairs.
    - The subsequent redox reactions on the semiconductor give off the reduced hydrocarbon products, and the excited electrons in the CB reduce CO_2, whereas holes in the VB oxidize water to oxygen. CO_2 reduction can occur via electron transfer and is associated with proton transfer. In contrast, it may also happen via proton-coupled

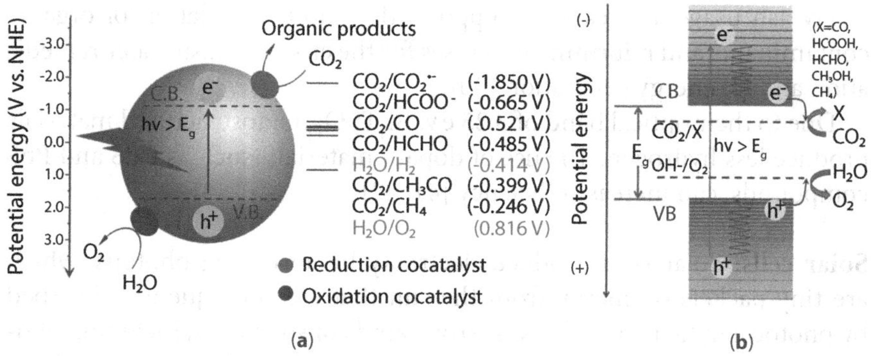

Figure 6.3 (a) Schematic illustration of a mechanism for CO_2 degradation and (b) relative energy levels of photocatalytic reduction of CO_2 on a semiconductor photocatalyst (https://pubs.acs.org/doi/10.1021/acsomega.7b00721).

electron transfer (PCET) to give off carboxylic acids, alcohols, methane, etc., where CH_4 is most thermodynamically preferred compared with other reduced products. The various possible reactions are shown in Figure 6.3.

The PCET reactions' reduction potentials are in the range of 0.7 V to 0.2 V (Figure 6.3), which is comparable to water's reduction potential (0.414 V at pH 7). Water reduction is a two-electron process, in contrast to the multielectron transfer needed to reduce CO_2 products (HCHO, CH_3OH, and CH_4) [44]. Water reduction subsequently competes with CO_2 reduction.
Methods of reduction include the following:

- Semiconductor-based photocatalysts like TiO_2, ZnO, CdS, and SiC are used in aqueous suspension form to reduce CO_2.
- The reduction reactions are catalyzed by co-catalysts like Ag, Au, Pt, and Pd after they receive electrons from the photocatalyst. Because of its poor binding to CO, Ag exhibits higher CO_2 reduction activity.
- The bandgap between the VB and CB is diminished by doping or co-doping. It helps to use more readily available visible light rather than expensive UV radiation sources. For instance, doping N as in CuO/ $TiO_{2-x}N_x$ increases the photoreduction of CO_2 to CH_4 or Ce in TiO_2 and exhibits increased activity compared to undoped TiO_2. It also exhibits a bathochromic shift in the absorption onset with visible light absorption.

- Nowadays, multicomponent photocatalysts are in use due to their low bandgap and ease of operation with visible light. Examples include TiO_2–RuH, an oxo-bridged complex for higher reduction of CO_2 into CH_4.
- Semiconductor MOFs (metal–organic frameworks), composites of graphitic C_3N_4, etc., are some other photocatalytic pathways to reduce CO_2 [45].

- **Wastewater treatment:** Chemicals are used in almost every industry, from agriculture (pesticides, herbicides, and insecticides) to large factories (which use a broad range of chemicals). These chemicals are consequently drained down in water bodies due to over-irrigation of fields or as factory effluents, and they increase the PBT (persistence, bioaccumulation, and toxicity) count of pollutants. To eliminate dyes and organic pollutants [10], including phenol, pharmaceuticals, pesticides, nitrogen, sulfur compounds, etc., photocatalytic degradation methods have been investigated.
- **Dyes:** Textile industries directly discharge about 1%–20% of their total production of dyes as effluents that are difficult to degrade by biological modes like anaerobic digestion by bacteria. Heterogeneous photocatalytic materials like TiO_2 and ZnO address this and can decolorize the dyes in minutes, with ZnO exhibiting superior activity to TiO_2 [11, 13, 14].
- **Phenols:** Typically, phenolic compounds are oxidized using TiO_2 and ZnO. However, investigations using TiO_2/g-C_3N_4 (TCN) thin-film electrodes showed that they completely oxidized phenol when used with a 1.5-V bias for 1.5 h under simulated sunlight [15]. La-doped ZnO degraded 2-chlorophenol from aquatic agriculture waste with a degradation efficiency of 83.92% with a catalytic dose of 10 mg and an irradiation time of 2 h at an optimum pH of 2 [46].
- **Nitrogen compounds:** The eutrophication potential and toxicity of nitrogenous substances make them a serious environmental issue. Recent experimental research demonstrates that using a photocatalyst made of phosphotogenic acid ($H_3PW_{12}O_{40}$) coated on a titanium dioxide (TiO_2) substrate degrades the pollutant by 94.1%. Additionally, research demonstrates that the La/Fe/TiO_2 catalyst has a higher activity for the optical degradation of ammonia nitrogen compared to the undoped pure TiO_2 [16, 47].

- **Pesticides:** Agricultural pollutants are incredibly harmful to the aquatic ecosystems and raise BOD and COD levels of water bodies, making the marine organisms' survival lethargic. Aqueous suspension of TiO_2 and $Na_2S_2O_8$ is utilized to combat this, which completely breaks down insecticides when exposed to sunlight [48].
- **Biomedical applications:** Photo catalytic materials have potential biomedical uses in the delivery of drugs, vaccination, cell imaging, gene therapy, early diagnosis and treatment of diseases, engineering of bone and cartilage tissues, medical devices, a biosensor for viruses and cancer molecules, cancer immunotherapy, wound healing, dental treatment, as well as antimicrobial and antioxidant agents [49, 50]. Moreover, photocatalytic cancer treatment is also documented [51, 52].
- **Supercapacitors:** In next-generation energy-storing systems, the supercapacitor is an effective energy storage device for high-power intensive operations. Electrode materials can be classified into two types of supercapacitors based on their mechanistic energy storage capabilities: (i) electrochemical double-layer and (ii) pseudocapacitors or redox capacitors, which use MnO, NiO, ZnO, RuO_2, CoO, etc. [53]. Due to their exceptional qualities, semiconductor oxides including SnO_2, TiO_2, NiO, and ZnO have attracted attention as electrode materials for supercapacitor applications. In terms of cost and environmental suitability, ZnO is the most readily available semiconductor oxide. Additionally, doping has improved their specific capacitance value and photodegradation efficiency [54, 55].

6.3.3 Photoelectrochemical Approach in Wastewater Treatment

Photoelectrochemical cells (PEC) have gained popularity due to their extensive and efficient utility in the treatment of wastewater and the production of valuable fuels like H_2, which is often regarded as a green and future fuel. PECs harvest light energy to drive chemical reactions by semiconductor material producing e/h^+ pairs that external electrical circuits can separate to carry out electrochemical reactions at respective electrodes. Each PEC consists of a photoelectrode primarily of photo-sensitive semiconductor material and can be photoexcited to liberate e^-/h^+ pairs and an auxiliary metal and reference electrode. PECs are akin to photovoltaic cells but operate differently as PECs are kept dipped in the electrolytic solution, and from within, they absorb photons to undergo electrochemical reactions.

AOP can be performed individually or in combination. AOPs generate free OH radicals primarily responsible for degrading organic pollutants in wastewater. AOPs generally employ two fundamental processes:

(I) Electrochemical conversion entails the direct or indirect photoelectron transfer (PET) conversion of organic contaminants into simpler biodegradable compounds.
(II) Electrochemical combustion: This process uses generated free radicals to mineralize organic molecules.

The Energy Policy Act of 1992 classifies hydrogen as an alternative fuel. Hydrogen's potential as a domestic alternative fuel source, capacity to power fuel cells in zero-emission vehicles, quick filling time, and high efficiency all contribute to its appeal as a transportation fuel. So, one of the most challenging research areas in recent years has been the creation of H_2 gas from wastewater.

Cho *et al.* (2021) showed the erbium-doped TiO_2 nanotubes [56] for the biodegradation of dyes like methylene blue, one of the sources of organic pollutants discharged by the textile and other chemical industries. The methylene blue may be degraded under the light in around 180 min with a photocatalytic efficiency of 85%. The efficiency of H_2 production was observed to be 17.39 mM h^{-1} cm^{-2} [57]. The use of CNT–Cd-ZnS/Fe2O3–based nanotubes by Anthony Raja and Preethi in 2020 [58] showed that they have the potential to curb and degrade highly toxic sulfide-containing effluents with the most excellent H_2 production being around 2679 mmol/h. A similar process was used by Vikrant *et al.* (2019) to produce hydride and 80 kJ/mol of energy by decomposing highly toxic hydrogen sulfide gas (H_2S) from wastewater [59].

Chavan and Fulekar, in 2018, combined photocatalyst oxidation with biodegradation to treat CEPT effluent [60]. Activated carbon (AC), bio-film, and TiO_2 were all employed in the reactor. The outcome revealed that wastewater was freed from TSS by 88%, ammoniacal nitrogen by 50%, COD by 61%, BOD by 92%, and decolorization by 78% in sunlight, TiO_2/AC, and bio-film.

Photoelectrochemical cells (PECs) used in treating wastewater and producing valuable fuels like H_2, which are widely considered as green and future fuel, have increased due to their extensive and efficient application in these fields.

6.4 Conclusion

Fossil fuels, such as coal, oil, and natural gas, are the basis of modern life, fueling every single thing from our homes' electrical power to the vehicles we drive to work. Three negative effects of burning fossil fuels include air pollution, water contamination, and climate change. The substances produced while burning fossil fuels are to blame for these bad impacts. In electrical terms, hydrogen has a higher energy density (33.6 kWh/kg) than diesel, which only retains about 12–14 kWh/kg of usable energy. Zero harmful emissions are also an essential characteristic of hydrogen gas. We can employ highly energetic hydrogen to reduce these negative impacts. Yet once more, to produce hydrogen, we need energy, which requires fossil fuels. We have no choice but to create green hydrogen using light energy as a source. Solar-driven hydrogen generation from water is a potentially effective solution to the environmental issues and the current global energy dilemma associated with fuel manufacturing.

Thus, the demand for a rapid and efficient photocatalytic hydrogen production method is significant. The potential of wastewater as a source for energy recovery has been discussed in this chapter, employing photocatalytic oxidation of contaminants in conjunction with hydrogen production. Energy-wise, synthesizing hydrogen from waste and pollutants is more advantageous than creating hydrogen from water splitting. The most widely used method to date has been suspensions of photocatalytic particles, while PECs have only been used in a few studies. PEC is a promising alternative because it lowers the system's recombination losses. Yet, the creation of hydrogen associated with the breakdown of contaminants in wastewater is a fascinating field for study and may have the potential for scaling up, at least in specific applications. We hope the young researchers will be fascinated by this chapter and investigate the diverse perspectives for producing hydrogen, reducing carbon dioxide, and degrading pollutants.

6.5 Challenges and Future Aspects

Undoubtedly, the production of hydrogen using photocatalysis is a very promising alternative to fossil fuels, but several issues arise because of technical limitations and other issues. Wastewater treatment, one of the primary applications of photocatalysis other than biomedical and storage applications [61, 62], requires semiconductor oxides [63] with wide applicability that may be employed in varying pH, temperature, and chemical composition of wastewater in varied environments to remove restrictions on operating parameters. In photoreactors, separation and unit washing processes that would normally cause concerns such as photocatalyst leaching and other issues can be prevented by catalyst immobilization. Prevention of electron–hole combination in photo-materials is a crucial factor to take into account as doing so could reduce the QY of the electronic process, which, in turn, reduces the activity of the photocatalyst. The scope of doping is also crucial to increase the photocatalytic activity [64, 65]. The generation of electron–hole pairs is the fundamental step during photoexcitation in a catalyst, which is responsible for the subsequent reduction of effluents to benign compounds less harmful to the environment. The well-established industrial application of photocatalysis has sparked a requirement for photocatalytic self-cleaning surfaces. Other than this, the photocatalysis phenomena have not yet been used extensively. The continually rising global energy demand is the biggest need to be addressed in the 21st century. The use of PEC water splitting for generating solar hydrogen appears to be an appealing and long-term solution. There are a few challenges, like robust photoelectrodes capable of harnessing the entire solar spectrum, that need to be resolved before the widespread use of PEC technology is feasible.

References

1. Jing, Z., Liu, G., Hang, G., Lian, L., Shi-huai, D., A theoretical basis for the relationship between industrial pollutant generation, abatement, emission, and economy. *Clean Technol. Environ. Policy*, 15, 707–711, 2013.
2. Tibbetts, J., Water world 2000. *Environ. Health Perspect.*, 108, 2, 69–73, 2000.
3. Guillebaud, J. and Hayes, P., Population growth and climate change. *BMJ*, 337, 247-248, 2008.
4. Kumar, R., Qureshi, M., Vishwakarma, D., Al-Ansari, N., Kuriqi, A., Elbeltagi, A., Saraswat, A., A review of emerging water contaminants and the application of sustainable removal technologies. *Case Stud. Chem. Environ. Eng.*, 6, 100219, 2022.
5. Long, Y., Xu, J., Shen, D., Du, Y., Feng, H., Effective removal of contaminants in landfill leachate membrane concentrates by coagulation. *Chemosphere*, 167, 512–519, 2017.

6. Huang, Y. and Koseoglu, S., Separation of heavy metals from industrial waste streams by membrane separation technology. *Waste Manag.*, 13, 5-7, 481–501, 1993.
7. Sahoo, T. and Prelot, B., Adsorption processes for the removal of contaminants from wastewater: the perspective role of nanomaterials and nanotechnology, in: *Nanomaterials for the detection and removal of wastewater pollutants*, pp. 161–222, Elsevier: Amsterdam, Netherlands, 2020.
8. Bhatia, D., Sharma, N., Singh, J., Kanwar, R., Biological methods for textile dye removal from wastewater: A review. *Crit. Rev. Environ. Sci. Technol.*, 47, 19, 1836–1876, 2017.
9. Kato, H., Asakura, K., Kudo, A., Highly efficient water splitting into H_2 and O_2 over lanthanum-doped $NaTaO_3$ photocatalysts with high crystallinity and surface nanostructure. *J. Am. Chem. Soc*, 125, 10, 3082–3089, 2003.
10. Han, F., Kambala, V., Srinivasan, M., Rajarathnam, D., Naidu, R., Tailored titanium dioxide photocatalysts for the degradation of organic dyes in wastewater treatment: a review. *Appl. Catal. A*, 359, 1-2, 25–40, 2009.
11. Houas, A., Lachheb, H., Ksibi, M., Elaloui, E., Guillard, C., Herrmann, J., Photocatalytic degradation pathway of methylene blue in water. *Appl. Catal. B: Environ.*, 31, 2, 145–157, 2001.
12. Foglia, L., Vempati, S., Tanda, B., Gierster, L., Wolf, M., Sadofev, S., Stähler, J., Revealing the competing contributions of charge carriers, excitons, and defects to the non-equilibrium optical properties of ZnO. *Struct. Dyn.*, 6, 3, 034501, 2019.
13. Arslan, I. and Balcioğlu, I.A., Degradation of commercial reactive dyestuffs by heterogenous and homogenous advanced oxidation processes: a comparative study. *Dyes Pigm.*, 43, 2, 95–108, 1999.
14. Weber, E. and Adams, R., Chemical-and sediment-mediated reduction of the azo dye disperse blue 79. *Environ. Sci. Technol.*, 29, 5, 1163–1170, 1995.
15. Ahmed, S., Rasul, M., Martens, W., Brown, R., Hashib, M., Heterogeneous photocatalytic degradation of phenols in wastewater: a review on current status and developments. *Desalination*, 261, 1-2, 3–18, 2010.
16. Qi, X., Zhuang, Y., Yuan, Y., Gu, W., Decomposition of aniline in supercritical water. *J. Hazard. Mater.*, 90, 1, 51–62, 2002.
17. Ameta, S. and Ameta, R. (Eds.), *Advanced oxidation processes for wastewater treatment: emerging green chemical technology*, Academic press: London EC2Y 5AS, United Kingdom, 2018.
18. Hamdany, A., Satyanaga, A., Zhang, D., Kim, Y., Kim, J., Photocatalytic cementitious material for eco-efficient construction—A systematic literature review. *Appl. Sci.*, 12, 17, 8741, 2022.
19. Rogé, V., Bahlawane, N., Lamblin, G., Fechete, I., Garin, F., Dinia, A., Lenoble, D., Improvement of the photocatalytic degradation property of atomic layer deposited ZnO thin films: the interplay between film properties and functional performances. *J. Mater. Chem. A*, 3, 21, 11453–11461, 2015.

20. Ebrahiem, E., Al-Maghrabi, M., Mobarki, A., Removal of organic pollutants from industrial wastewater by applying photo-Fenton oxidation technology. *Arabian J. Chem.*, 10, 1674–1679, 2017.
21. Ahile, U., Wuana, R., Itodo, A., Sha'Ato, R., Malvestiti, J., Dantas, R., Are iron chelates suitable to perform photo-Fenton at neutral pH for secondary effluent treatment? *J. Environ. Manag.*, 278, 111566, 2021.
22. Kudo, A. and Miseki, Y., Heterogeneous photocatalyst materials for water splitting. *Chem. Soc Rev.*, 38, 1, 253–278, 2009.
23. Pardeshi, S. and Patil, A., Solar photocatalytic degradation of resorcinol a model endocrine disrupter in water using zinc oxide. *J. Hazard. Mater.*, 163, 1, 403–409, 2009.
24. Wei, Z., Liang, F., Liu, Y., Luo, W., Wang, J., Yao, W., Zhu, Y., Photoelectrocatalytic degradation of phenol-containing wastewater by $TiO_2/g\text{-}C_3N_4$ hybrid heterostructure thin film. *Appl. Catal. B: Environ.*, 201, 600–606, 2017.
25. Lam, S., Sin, J., Abdullah, A., Mohamed, A., Photocatalytic degradation of resorcinol, an endocrine disrupter, by TiO2 and ZnO suspensions. *Environ. Technol.*, 34, 9, 1097–1106, 2013.
26. Nuraje, N., Asmatulu, R., Kudaibergenov, S., Metal oxide-based functional materials for solar energy conversion: a review. *Curr. Inorg. Chem.*, 2, 2, 124–146, 2012.
27. Poungchan, G., Ksapabutr, B., Panapoy, M., One-step synthesis of flower-like carbon-doped ZrO_2 for visible-light-responsive photocatalyst. *Mater. Des.*, 89, 137–145, 2016.
28. Sirajudheen, P., Poovathumkuzhi, N., Vigneshwaran, S., Chelaveettil, B., Meenakshi, S., Applications of chitin and chitosan-based biomaterials for the adsorptive removal of textile dyes from water—A comprehensive review. *Carbohydr. Polym.*, 273, 118604, 2021.
29. Wang, B., Wan, Y., Zheng, Y., Lee, X., Liu, T., Yu, Z., Huang, J., Ok, Y.S., Chen, J., Gao, B., Alginate-based composites for environmental applications: a critical review. *Crit. Rev. Environ. Sci. Technol.*, 49, 4, 318–356, 2019.
30. Hamidon, T., Adnan, R., Haafiz, M., Hussin, M., Cellulose-based beads for the adsorptive removal of wastewater effluents: a review. *Environ. Chem. Lett.*, 20, 3, 1965–2017, 2022.
31. Nasrollahzadeh, M., Sajjadi, M., Iravani, S., Varma, R., Starch, cellulose, pectin, gum, alginate, chitin and chitosan derived (nano) materials for sustainable water treatment: A review. *Carbohydr. Polym.*, 251, 116986, 2021.
32. Kumar, A., Sharma, G., Naushad, M., Ala'a, H., García-Peñas, A., Mola, G., Stadler, F., Bio-inspired and biomaterials-based hybrid photocatalysts for environmental detoxification: A review. *Chem. Eng. J.*, 382, 122937, 2020.
33. Maeda, K. and Domen, K., Photocatalytic water splitting: recent progress and future challenges. *J. Phys. Chem. Lett.*, 1, 18, 2655–2661, 2010.
34. Del, F., Ishikawa, A., Domen, K., de La Mano, J., Sánchez-Sánchez, M., González, I., Herreras, S., Mota, N., Rivas, M., Galván, M., Fierro, J., Influence

of Zn concentration in the activity of Cd1− xZnxS solid solutions for water splitting under visible light. *Catal. Today*, 143, 1-2, 51–56, 2009.
35. Nagabhushana, G., Nagaraju, G., Chandrappa, G., Synthesis of bismuth vanadate: its application in H_2 evolution and sunlight-driven photodegradation. *J. Mater. Chem. A*, 1, 2, 388–394, 2013.
36. Koli, P., Shinde, S., Kapadnis, K., Patil, A., Shinde, M., Khairnar, S., Sonawane, D., Ingale, R., Transition metal incorporated, modified bismuth oxide (Bi2O3) nano photo catalyst for deterioration of rosaniline hydrochloride dye as resource for environmental rehabilitation. *J. Indian Chem. Soc.*, 98, 11, 100225, 2021.
37. Sobczynski, A., Yildiz, A., Bard, A., Campion, A., Fox, M., Mallouk, T., Webber, S., White, J., Tungsten disulfide: a novel hydrogen evolution catalyst for water decomposition. *J. Phys. Chem.*, 92, 8, 2311–2315, 1988.
38. Jin, L., Lv, S., Miao, Y., Liu, D., Song, F., Recent development of porous porphyrin-based nanomaterials for photocatalysis. *ChemCatChem*, 13, 1, 140–152, 2021.
39. Godemann, C., Hollmann, D., Kessler, M., Jiao, H., Spannenberg, A., Brückner, A., Beweries, T., A model of a closed cycle of water splitting using ansa-titanocene (III/IV) triflate complexes. *J. Am. Chem. Soc.*, 137, 51, 16187–16195, 2015.
40. Nguyen, H., Nguyen, H., Ahmed, S., Rajamohan, N., Yusuf, M., Sharma, A., Arunkumar, P., Deepanraj, B., Tran, H., Al-Gheethi, A., Vo, D., Emerging waste-to-wealth applications of fly ash for environmental remediation: A review. *Environ. Res.*, 227, 115800, 2023.
41. Gahlawat, A., Sindhu, M., Sharma, A., Maan, K.S., Singh, P.P., Aepuru, R., Kumar, D., Synthesis and characterization of Bismuth titanate perovskite materials for hydrogen production. *Mater. Lett.*, 342, 134316, 2023.
42. Amika, G., Sindhu, M., Sharma, A., Maan, K., Singh, P., Ladiya, S., Kumar, D., Sharma, R., Dahiya, P., Facile Synthesis of CdS Doped Bismuth Titanate Perovskites: A Cocatalyst-Free Promising Photocatalyst for Hydrogen Generation. *Chem. Afr.*, 6, 3109–3117, 2023.
43. Boskabadi, M., Rogé, V., Bazargan, A., Sargazi, H., Barborini, E., *An introduction to photocatalysis. Photocatalytic Water and Wastewater Treatment*, pp. 1–36, IWA Publishing: London, UK, 2022.
44. Takanabe, K., Photocatalytic water splitting: quantitative approaches toward photocatalyst by design. *ACS Catalysis*, 7, 11, 8006–8022, 2017.
45. Buettner, G., The pecking order of free radicals and antioxidants: lipid peroxidation, α-tocopherol, and ascorbate. *Arch. Biochem. Biophys.*, 300, 2, 535–543, 1993.
46. Rani, S., Sharma, A., Tabasum, S., Malik, A., Chaudhary, S., Kumar, D., Singh, H., Singh, P., Highly Efficient Photocatalytic Properties of La-Doped ZnO over Pristine ZnO for Degradation of 2-Chlorophenol from Aquatic Agriculture Waste. *Chem. Afr.*, 6, 1–10, 2023.

47. Nikokavoura, A. and Trapalis, C., Graphene and g-C_3N_4 based photocatalysts for NOx removal: a review. *Appl. Surf. Sci.*, 430, 18–52, 2018.
48. Wang, W., Huang, Y., Yang, S., Photocatalytic degradation of nitrobenzene wastewater with H3PW12O40/TiO2, in: *2010 International Conference on Mechanic Automation and Control Engineering*, IEEE, pp. 1303–1305, 2010.
49. Luo, X., Chen, C., Yang, J., Wang, J., Yan, Q., Shi, H., Wang, C., Characterization of La/Fe/TiO2 and its photocatalytic performance in ammonia nitrogen wastewater. *Int. J. Environ. Res. Public Health*, 12, 11, 14626–14639, 2015.
50. da Silva, W., Druzian, D., Oviedo, L., Muraro, P., Oviedo, V., Silver Nanoparticles for Photocatalysis and Biomedical Applications, in: *Silver Micro-Nanoparticles-Properties, Synthesis, Characterization, and Applications*, IntechOpen: London, UK, 2021.
51. Zhao, B., Wang, Y., Yao, X., Chen, D., Fan, M., Jin, Z., He, Q., Photocatalysis-mediated drug-free sustainable cancer therapy using nanocatalyst. *Nat. Commun.*, 12, 1, 1345, 2021.
52. Abdelwaheb, S, Photocatalytic metal bionanocomposites for biomedical applications, in: *Micro and Nano Technologies, Bionanotechnology : Emerging Applications of Bionanomaterials*, Barhoum, A., Jeevanandam, J., Danquah, M.K. (Eds.), pp. 227–258, 2022.
53. Huang, T., Qiu, Z., Wu, D., Hu, Z., Bamboo-based activated carbon-MnO_2 nanocomposites for flexible high-performance supercapacitor electrode materials. *Int. J. Electrochem. Sci.*, 10, 8, 6312–6323, 2015.
54. Uke, S., Mardikar, S., Bambole, D., Kumar, Y., Chaudhari, G., Sol-gel citrate synthesized Zn doped $MgFe_2O_4$ nanocrystals: a promising supercapacitor electrode material. *Mater. Sci. Energy Technol.*, 3, 446–455, 2020.
55. Blake, D., Maness, P., Huang, Z., Wolfrum, E., Huang, J., Jacoby, W., Application of the photocatalytic chemistry of titanium dioxide to disinfection and the killing of cancer cells. *Sep. Purif. Methods*, 28, 1, 1–50, 1999.
56. Sieben, J., Morallón, E., Cazorla, D., Flexible ruthenium oxide-activated carbon cloth composites prepared by simple electrodeposition methods. *Energy*, 58, 519–526, 2013.
57. Cho, H., Joo, H., Kim, H., Kim, J., Kang, K., Yoon, J., Improved photoelectrochemical properties of TiO_2 nanotubes doped with Er and effects on hydrogen production from water splitting. *Chemosprahere*, 267, 129289, 2021.
58. Raja, M. and Preethi, V., Photocatalytic hydrogen production using bench-scale trapezoidal photocatalytic reactor. *Int. J. Hydrogen Energy*, 45, 13, 7574–7583, 2020.
59. Vikrant, K., Kim, K., Deep, A., Photocatalytic mineralization of hydrogen sulfide as a dual-phase technique for hydrogen production and environmental remediation. *Appl. Catal. B: Environ.*, 259, 118025, 2019.
60. Chavan, A. and Fulekar, M., Integration of photocatalytic oxidation and biodegradation treatment processes to enhance degradation efficiency of CETP wastewater contaminants. *Bionanosci.*, 8, 761–768, 2018.

61. Shi, C., Kang, F., Zhu, Y., Teng, M., Shi, J., Qi, H., Huang, Z., Si, C., Jiang, F., Hu, J., Photoreforming lignocellulosic biomass for hydrogen production: Optimized design of photocatalyst and photocatalytic system. *Chem. Eng. J.*, 452, 138980, 2023.
62. Su, R., Tiruvalam, R., Logsdail, A., He, Q., Downing, C., Jensen, M., Dimitratos, N., Kesavan, L., Wells, P., Bechstein, R., Jensen, H., Designer titania-supported Au–Pd nanoparticles for efficient photocatalytic hydrogen production. *ACS Nano*, 8, 4, 3490–3497, 2014.
63. Ni, M., Leung, M., Leung, D., Sumathy, K., A review and recent developments in photocatalytic water-splitting using TiO2 for hydrogen production. *Renew. Sustain. Energy Rev.*, 11, 3, 401–425, 2007.
64. Afsari, M., Youzbashi, A., Nuranian, H., Zahraee, S., Remarkable improvement of visible light photocatalytic activity of TiO2 nanotubes doped sequentially with noble metals for removing of organic and microbial pollutants. *Mater. Res. Bull.*, 94, 15–21, 2017.
65. Loganathan, K., Bommusamy, P., Muthaiahpillai, P., Velayutham, M., The syntheses, characterizations, and photocatalytic activities of silver, platinum, and gold doped TiO_2 nanoparticles. *Environ. Eng. Res.*, 16, 2, 81–90, 2011.

7

Recent Advances on the Electrode Materials Used in Microbial Fuel Cell for Simultaneous Power Generation and Wastewater Treatment

Ankit Kumar, Soumya Pandit*, Shikha Singh and Chetan Pandit

Department of Life Sciences, School of Basic Sciences and Research, Sharda University, Greater Noida, India

Abstract

Microbial fuel cell (MFC) is an advanced technology that is used for producing electricity or energy from waste material with the help of microbes, which would be commercialized in the future. In MFC, bacteria oxidize organic/inorganic material and transform chemical energy into electrical energy. It is an advanced technique that has the potential to serve as a major renewable source of energy. In the past few years, many different kinds of electrode materials (anode and cathode, such as carbon and CNT-based materials) and catalysts (improve the performance of the biochemical cell reaction such as microbes) have been examined for possible improved outcomes of MFC. The efficiency of the MFC is effectively determined by the electrode material used in it, and several electrode surface treatments are done. In this review, advances in suitable electrode (cathode) material that can be used to enhance the productivity of MFC are systematically discussed.

***Keywords*:** Anode, microbial fuel cell, power density, nanomaterials, electroactive bacteria

**Corresponding author*: sounip@gmail.com

Sadanand Pandey, Elvis Fosso-Kankeu and Soumya Pandit (eds.) Synergy of Bio-Chemical Processes for Photocatalytic and Photoelectrochemical Wastewater Treatment, (161–188) © 2024 Scrivener Publishing LLC

7.1 Introduction

In the present scenario, electricity demand is highly increased because of the increased population and industrialization. To meet this demand, electricity is being produced at electricity-generating plants mostly from fossil fuels and nuclear power. For generating electricity, fossil fuels are being buried, which results in the emission of carbon dioxide at a higher level causing pollution. Due to pollution, today, the whole world is facing major environmental issues such as climate change and global warming. To overcome such environmental issues simultaneously, to meet the increased electricity demand, there is a need to find the best eco-friendly alternative [1]. Electricity production techniques possessing the potential to generate electricity from renewable sources and operating an eco-friendly approach that are sophisticated at low cost need to be developed.

A British botanist Potter has developed an idea in 1911, the production of electricity using microbes, which oxidizes organic molecules and converts chemical energy into electricity [2]; his discovery attracted the attention of researchers worldwide. Microbial fuel cell (MFC) is an advanced technique that has the potential to produce energy from renewable sources fulfilling an eco-friendly approach. Potter initially came up with the concept in 1911; however, it was not accepted by many till 1999, as it was demonstrated that bacteria are capable of transferring electrons externally to electrodes. The process has been refined over the last 10 years in a way that renders it practicable for both wastewater treatment and electricity generation.

In the future, there will be less availability of clean and drinkable water due to water pollution. To meet this high demand for clean and drinkable water, MFC is one of the promising solutions for wastewater treatment; MFC utilizes different carbohydrates and complex substances (organic/inorganic) from wastewater producing electricity, simultaneously providing clean water by degrading pollutants.

But MFC is not well commercialized yet for electricity production due to its limitations such as low energy generation ability and high cost. Over the last decade, various research studies on MFC have been performed and concluded that MFC efficiency can be enhanced by improving microorganism adhesion to electrode material and electron transfer. It is found that microorganisms perform different metabolic pathways generating an efficient source of energy. Adhesion of microorganisms to electrode material can be enhanced using suitable modified electrode material. However the major challenge is to develop efficient electrode material at a cheap cost.

The material used for the electrode has been the primary focus of this field's study for the last 10 years because it greatly impacts the performance of MFC, thus affecting their potential for commercialization. Mostly, electrode materials used in MFCs are carbon, as most suitable because of its nature of not being corrosive and the general characteristics of an electrode. Currently, across the world, researchers are working to improve the efficiency of MFC by developing suitable modified electrode materials.

7.2 Introduction to MFC

Microbial fuel cell (MFC) (Figure 7.1) is an electrochemical fuel cell in which waste material (organic) is oxidized by microorganisms to form electricity. MFC transforms chemical energy into electricity without burning and releases carbon dioxide into an external environment. Thus, it is an eco-friendly approach for electricity generation. In MFC, electricity is produced by substrate oxidation with the help of microbes. During this process, oxidation-reduction reactions are catalyzed at the electrode due to the microbial extracellular electron exchange process, which results in the release of electrons. This is known as electrocatalysis.

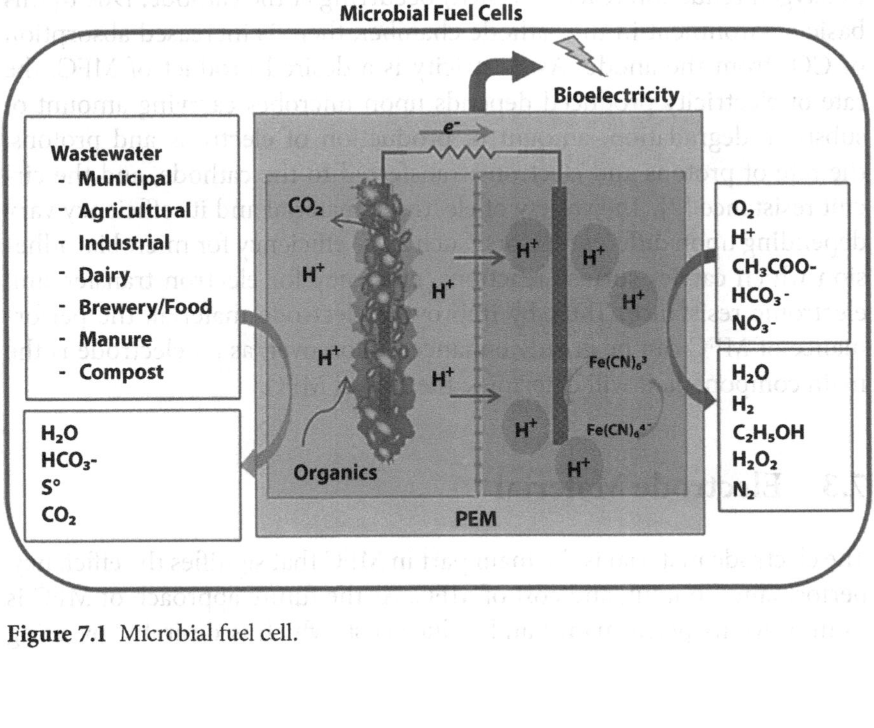

Figure 7.1 Microbial fuel cell.

Microbial fuel cell (MFC) contains electrodes microbes, electrolyte material (connected by both electrodes), and proton interchange membrane such as Nafion and PTFE (polytetrafluoroethylene), which separates anode and cathode chamber from each other. MFC utilizes waste material (organic/inorganic) to generate energy by degrading substrates [3]. The gaseous end product CO_2 and acetate, glucose, or waste material is used as substrate in an anode chamber [4].

Microbes in an anode chamber oxidize organic substances (substrate) under the anaerobic condition to produce CO_2, electrons, and protons. Electrons produced by microbe migrate towards anode by direct-membrane accessory electron transfer. Electron mediators or nanowires carry electrons from anode to cathode, which are connected to an external circuit unit [5], thereby producing electricity as the main product. In addition, protons from the anode chamber migrate to the cathode chamber through PEM. Electrons and protons are later combined in the cathode chamber, causing a reduction reaction at the cathode by reducing oxygen to water. Bacterial biofilm is produced at the anode by microorganisms and catalyzes the conversion of chemical energy from substrates into electrons, while O_2 is reduced to H_2O at the cathode [6].

The cathode chamber has an oxygen inlet that facilitates reduction reaction, giving H_2O as an end product. In the cathode chamber, the alkaline condition is produced because of aggregation of the hydroxide ions due to oxygen reduction reaction (ORR) occurring at the cathode. Due to this basic environment in the cathode chamber, there is increased absorption of CO_2 from the anode. As electricity is a desired product of MFC, the rate of electricity produced depends upon microbes carrying amount of substrate degradation, amount of production of electrons and protons, the rate of protons and electrons transferred to the cathode, and the circuit resistance [7]. The variety of electrode material and its efficiency vary depending upon different factors. Such as its efficiency for microbial adhesion which carries surface reactions, efficiency for electron transfer, and electronic resistance. Thus, by improving electrode material, the performance of MFC can be greatly enhanced. Moreover, as an electrode is the main component, it will determine the cost of MFC.

7.3 Electrode Material

The electrode material is the main part in MFC that signifies the efficiency, performance, output, and cost of MFC. As the future approach of MFC is to improve its performance and reduce cost, which is desirable by using suitable improved electrode material. The improved electrode material is meant to improve in its properties such as a large surface area for microbial adhesion, electron transfer efficiency, good compatibility, high electronic conductivity, and relatively low cost. In addition, according to Logan, the cathode material should possess catalytic properties for oxygen reduction to enhance MFC output.Many studies reveal that carbon maddens materials (such as carbon cloth, sheet, felt, foam, and paper) are more suitable electrode material because of it has many properties like as high conductivity of electron and porosity, good biocompatibility, better chemical stability, and relatively cheaper. But for implementing this project in practice at a large scale, there is a need for a technique that produces increased output enough to meet energy demand at a very low cost. To select suitable electrode material, there are different criteria for both anode and cathode. But there are some properties that should possess both. These are surface area, conductivity, stability and durability, and low cost and accessibility.

7.4 Anode Material

Anode plays a significant role in microbial adhesion and the rate of electron transfer from biomass to anode and anode to cathode, which of course decides the fate of MFC. Anode provides a surface for biofilm formation (microbial adhesion). To gain maximum electron production by microbes, an anode provides a platform to conduct various microbial reactions. As more surface area is provided, more will be the sites available for microbial reaction at the anode. As well as, the better the suitable anode material used, the better it will facilitate microbes for attachment. Various materials can be utilized for anode such as carbon-based materials, graphite, stainless steel, ceramic electrodes, and CNT-based anode material, which are discussed below.

7.4.1 The Traditional Carbon-Based Anode Material

Traditional carbon materials are widely used anode materials because of their biocompatibility, performance, and relatively low cost.

A graphite rod is a widely used electrode material because of its features such as chemical stability (sp2 hybridization) and high electrical conductivity. Liu *et al.* [8] worked on the use of graphite rods in MFC. They developed a single-chambered MFC that contained eight graphite rod anodes, a single air cathode, and sewage from the primary sedimentation tank as substrate. They observed the maximum power obtained was about 26 mW m^{-2} with 80% decrease in chemical oxygen demand (COD). But the applications of graphite rods were bounded because of their truncated porosity and surface area. Chaudhuri and co-workers overcome this problem using graphite felt anode instead of graphite rods [9]. They used graphite felt bioanodes in glucose-fed MFC containing *Rhodoferaxferrireducens* as substrate and observed three times more electricity produced than the graphite rod. He observed that the increased surface area facilitated the formation of dense biofilm that oxidizes a higher amount of glucose.

Ahn and Logan [10] developed a one-chamber air cathode MFC that involved graphite fiber brush anode wound surrounding the titanium wire that facilitates increased surface area, low electrode resistance, and higher microbial adhesion. They observed that the power density generated was 422 mW m^{-2}. In another work, Logan *et al.* [11] used graphite brush anode and [cobalt tetra methoxyphenyl porphyrin (CoTMPP)] as a cathode in a flat-type MFC and observed power density obtained of about 2,400 mW m^{-2}, and it was four times more than the carbon paper used.

A presentation by Wang *et al.* used a carbon cloth as anode with brewery wastewater as substrate in purposely built single-chambered MFC and obtained the electricity of 483 mW m^{-2}. He *et al.* [12] used a three-dimensional (3D), ultra-met, and chemically and thermally resistant reticulated vitreous carbon (RVC) material that was proven to be biocompatible. RVC is not much frequently used because of its higher resistance. He *et al.* got the current density of about 170 mW m^{-2} using RVC as the anode sample.

Kim *et al.* [13] developed a dual-chamber MFC containing carbon paper as anode, and the power obtained was 40 mW m^{-2}. Wang and his co-workers used carbon mesh as the anode that is cheaper than carbon cloth. They observed that the increased power density obtained was 1,015 mW m^{-2}, which was higher than the carbon cloth. Liang *et al.* [14] developed a new type of filler-type anode by sintering carbon felt and carbon paper, which enhances the connection between the filler and the carbon paper and reduces the anode resistance. They obtained an extremely high power density of 2,426 mW m^{-2}.

7.4.1.1 Granular Carbon

Granular carbon (GC), granular graphite (GG), and granular activated carbon (GAC) were the most used filling material used for anode in MFCs. These granules provide a larger surface area for biofilm adhesion and enhanced electric conductivity to the anode that greatly enhances the performance of MFCs. The electric contact with granules and external of the cell is used to maintain by graphite rods.

Firstly, Rabey *et al.* used GG stack as an anode in tubular-chamber MFC. The chamber was filled with 52% of GG and obtained power generation of up to 49 W/m^3. The GG material was used as anode in different modes like tubular chamber MFCs, single-chamber MFCs, double-chamber MFCs, and wetland-constructed MFC's. Later, Rabey *et al.* used tubular-chamber MFC containing GG. They used different substrates such as glucose, acetate, digester effluent, and household wastewater, and the power generation was observed between 48 W/m^3 and 90 W/m^3. When the wastewater was utilized as substrate, almost 96% of organic matter was consumed by microbes to generate electricity. This material was found to be a better alternative for wastewater treatment using MFCs.

He *et al.* [12] used GAC as an anode in the upflow mode of MFCs and produced electricity continuously for 5 months by feeding sucrose solution. The highest power density generated was around 170 mW/m^2. When the single-chamber MFC was run using GAC as anode material, the power generation was increased and adhesion of a higher amount of biomass was observed with more GAC granules added. It was found to be a better replacement for carbon cloth. MFC containing multianode-GAC was performed and found more electricity generated due to its better electrons conductivity and lower resistance. Wetland-constructed MFC was also built using GAC granules and tested for its performance. It was observed that the performance of packed bed GG or GAC was better and more compatible than carbon felt and carbon cloth in wetland MFCs. The GG with a diameter of 8–13 mm and initial porosity of 0.38 was used to construct wetland MFC. The power generated was about 276 mW/m^3 and was found to be a better performer in terms of the removal of pollutants.

Factors such as granular size, surface area, construction, and packing fraction influenced the performance of stacked granule carbon, whereas the granular size is the most. When the granules containing sizes between 0.25 mm and 0.5 mm were used, the current density obtained was 77.7 mA/m^2. But it fell to 37.9 mA/m^2 when the granular size used was between 1 mm and 5 mm. GG size containing 2 mm and 5 mm was used in double-chamber MFCs to compare its effect on the performance of MFCs. The GG with a size of 5 mm produced electricity of 257 mW/m^2, whereas the GG with a size of 2 mm generated electricity of 143 mW/m^2. The reason behind these might be due to more number of junction points between minute granules that influenced more resistance for electron transport.

The effect of packing fraction and surface area was studied using GG in single-chamber MFC. Whereas, it was found that a rise in thickness of anode bed and its surface area enhances the electricity output of MFC. Electricity generated about 1.3 W/m^3 was the best current output when the 3-cm anode bed was used. Coating with a mixture of quaternary DABCO polysulfone and carbon black (GC) enhanced the electrical conductivity of GG and found the solution resistance of ion exchange polymer-coated anodes to be more stable. To enhance the electrical conductivity and biofilm adhesion by modifying anode, GAC could be used to enhance modification treatment. In an experiment, GCA was used for a uniform flow field of electrons, and the current density obtained was enhanced up to 17%.

7.4.1.2 Stainless Steel

Stainless steel is a better alternative for carbon-based anodes because of its high mechanical strength, corrosion-resistant, and excellent electron conductivity ability. Due to these properties of stainless steel, it could be utilized for long-term high-scale MFC operations. In recent experiments, stainless steel has been proved a promising anode material for the efficiently stable current generation. In an experiment, a plain stainless steel anode was tested against a flat graphite electrode under similar experimental conditions. Results showed that the use of stainless steel benefited by more current density generation.

However, stainless steel was not that efficient to adhere thick film of bacteria, which may affect the current density production of MFC. To overcome this limitation stainless steel anode, Ketep *et al.* [15] produced a 3D stainless steel foam with appropriate porosity, containing a larger surface area. This anode could adhere higher amount of biomass and was more efficient to conduct increased electron transfer, which surely enhanced the performance of MFC. This 3D stainless steel foam produced a current density four times that of the plain stainless steel anode and about two times higher than the carbon cloth. MFCs with different configurations can be seen in Figure 7.2.

7.5 Carbon Nanotubes (CNTs) and Composite Anode Material

Carbon nanotubes (CNTs) have been proved to be a better substitute for anode material due to their good conductivity, biocompatibility, large surface area, and increased catalytic ability. Moreover, CNT facilitates increased cell adhesion and growth ability over anode surface.

It is reported that incorporation of CNTs into a conductive polymer leads to a synergistic effect. Qiao *et al.* [16] used CNT-polyaniline as anode

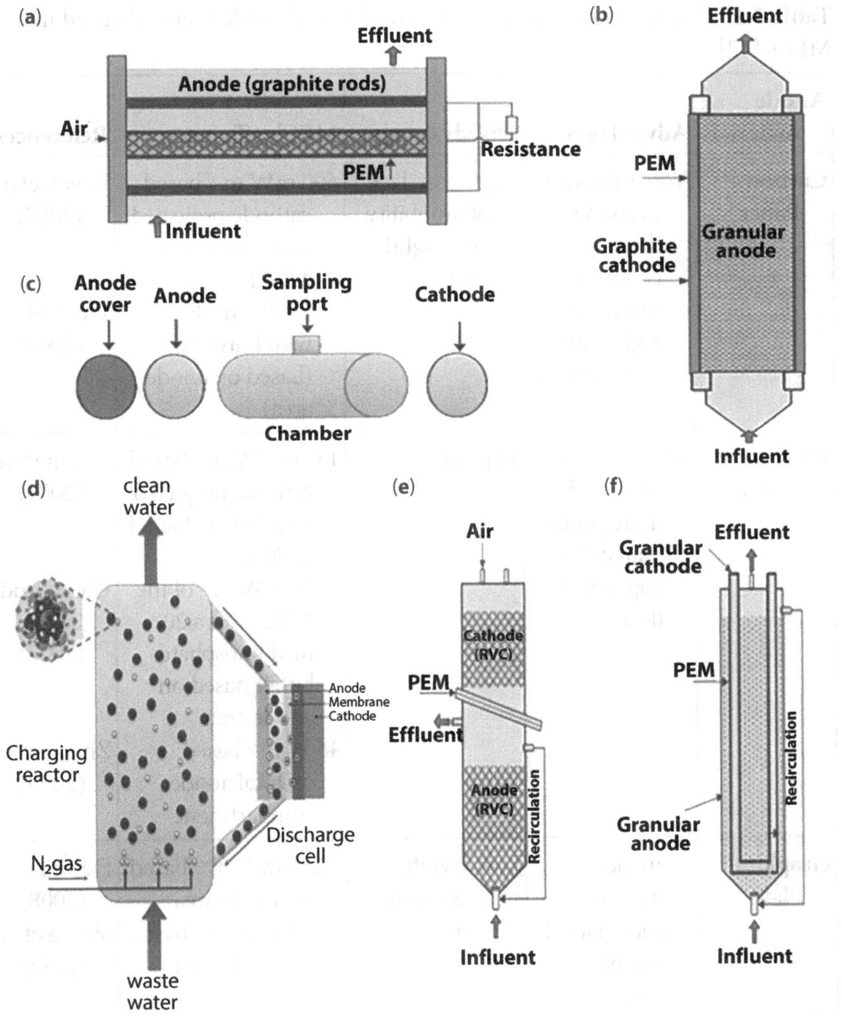

Figure 7.2 (a) The first SCMFC for domestic wastewater treatment [40]. (b) Tubular MFC (upflow mode of MFC) [40]; (c) A lab-scale single-chamber MFC [40]. (d) fluidized capacitive anode MFC [41]. (e, f) Upflow MFC with cylindrical shape [40].

material and reported that CNTs could enhance bacterial adhesion and electrocatalytic activity. They assumed that this provides not only a large surface area but also the protective effect of polyaniline. Zou *et al.* [7] found that modified carbon paper facilitates increased electrochemical properties when polypyrrole/CNTs were used as anode material. Conventional anode materials can be seen in Table 7.1.

Table 7.1 Characteristic comparison of traditional anode materials used in MFCs [42].

Anode materials	Advantages	Disadvantages	MFC's efficiency	References
Carbon Paper	Very thin, easy to connect the wire, compact structure, and relatively smooth area	High cost, lack of durability, and slightly brittle	600 mW m^{-2} based cathode projected area (bottle type MFC) 506 mW m^{-2} fed with butyrate (based on anode area)	Logan *et al.* (2007) Liu *et al.* (2005)
Carbon Cloth	More flexible and much more porous than carbon paper and thin	Expensive	1,040 mW m^{-2} based cathode projected area (cube-shaped MFCs) 1640 mW m^{-2} of the MFC with a 200 mM phosphate buffer based on anode area 46 W m^{-3} based on total anode compartment	Logan *et al.* (2007) Cheng and Logan (2007) Zhang *et al.* (2009)
Graphite Plate	Compact structure and relatively smooth surface	Low specific area and high cost	3,290 mW m^{-2} based on anode area 1,078 mW m^{-2} based on anode area	Dewan *et al.* (2008) Shewa *et al.* (2014)
Carbon Mesh	Less expensive alternative to carbon paper and carbon cloth, thin, and easy to deform		The data for untreated carbon mesh was not reported; the stable voltages were produced from MFCs with all anodes except the untreated carbon mesh (CM).	Wang *et al.* (2009)

(*Continued*)

Table 7.1 Characteristic comparison of traditional anode materials used in MFCs [42]. (*Continued*)

Anode materials	Advantages	Disadvantages	MFC's efficiency	References
Granular Graphite	High specific area	Porosities of the packed electrode are relatively low, clogging after long-term running	48 W m^{-3} and 38 W m^{-3} based on total anode compartment, for feed streams based on acetate and glucose, respectively	Rabey *et al.* (2005)
			175 W m^{-3} based on total anode compartment	Aelterman *et al.* (2008)
Graphite/ Carbon Felt	Relatively low cost and large porosity	Much thicker than then materials described above; the growth of bacteria is more likely to be restricted by the mass transfer of substrate and products on inner surface; and large resistance	356 W m^{-3} based on total anode compartment for carbon felt	Aelterman *et al.* (2008)
			365 W m^{-3} based on total anode compartment for graphite felt	Aelterman *et al.* (2008)
Reticulated Vitreous Carbon (RVC)	Available with different pore sizes, conductive and rigid	Brittle material	39.4 + 3.0 W m^{-3} based on cathode area	Lepage *et al.* (2012)
			170 mW m^{-2} based on anode area	He *et al.* (2007)

(*Continued*)

Table 7.1 Characteristic comparison of traditional anode materials used in MFCs [42]. (*Continued*)

Anode materials	Advantages	Disadvantages	MFC's efficiency	References
Carbon Brush	Ideal electrode with high porosities, and efficient current collection	Clumping of fibers hindered bacterial access to the fiber surfaces as well as the diffusion of substrate into the brush interior	2,400 Mw^{-2} based on cathode projected area (cube-shaped MFC) 1,430 mW m^{-2} based cathode projected area (bottle-type MFC)	Logan *et al.* (2007)

CNTs possess some properties, which found to be strongly beneficial to enhance the performance of MFCs, such as huge surface area (usually up to 1,300 $m^2 g^{-1}$) and improved catalytic properties. Furthermore, voids between the CNTs provide more area for microbes to grow. Wherever, Tsai *et al.* [17] modified the carbon cloth with CNTs providing a highly conductive anode with a large surface area that could improve the performance of MFC. They found that the power density obtained was increased by 250%. Sharma *et al.* [18] found that deposition of multiwall CNTs over carbon paper benefited MFC with power density obtained six times more than that with the graphite material.

Cui and his coworker played an excellent role in anode surface modification with CNTs. They constructed a 3D porous structured-CNT electrode capable of providing a better construction of anode for biofilm production and faster substrate movement in anode structure. Moreover, this anode surface modification with CNTs enhanced anode–biofilm interaction and electron transfer rate from biomass to the anode. To overcome the issue of conductivity of 3D CNT-structured anode, constructed 3D sponge coated with carbon nanoparticles showed enhanced electrical conductivity (approximately 1 S/cm), and the microporous CNT electrode and macroporous sponge together in combination produced up to 2.1 mA/cm^2 current density. MFCs show a high efficiency with CNTs, but the problem is with its complex fabrication and high cost, which limit its use at a higher scale.

7.5.1 Graphene

Graphene is a 2D lattice nanomaterial with sp^2-hybridization. Presently, the use of graphene as anode material vastly attracted researchers worldwide. It is a kind of honeycomb network containing excessive mechanical strength, excellent electron conductivity, increased surface area for biofilm adhesion, and stable chemical ability. Its long-range of Π-conjugation plate-shaped crumpled structure provides a broad range of applications for electricity production and its storage. Xiao *et al.* [19] obtained current density of 3.6 W/m^3 by using crumpled structure modified graphene anode. Its fabrication plays a vital role in MFC. It is demonstrated that utilization of this graphene material as an anode highly increases the performance and electricity output of MFC (almost doubled; from 1.7 mW/m^2 to 3.6 mW/m^2). Conventionally, carbon-based materials are being used for anode in MFCs. But the properties of graphene and its excellent results made it a better replacement for conventional carbon-based anode material (Table 7.2).

Firstly, Zhang *et al.* [20] introduced graphene as anode material for MFC. They used graphene-modified stainless steel material as an anode and observed current density achieved 18 times higher than the origin stainless steel mesh. The surface area offered by graphene for biofilm formation was 2,630 $m^2 g^{-1}$, which is large and carries electron mobility higher than 200,000 $cm^2 V^{-1} s^{-1}$. Deposition of graphene over a carbon cloth anode material showed the current density output elevated by 1.7 times from the origin. Utilization of this hybrid anode material highly enhanced electronic conductivity and biofilm formation over the anode. One researcher showed that the use of different dimensions of graphene highly enhances the properties of anode such as electron transfer efficiency and biofilm formation over the anode.

Fabricated *in situ* graphene structure on graphite electrode that was done using electrochemical exfoliation treatment was experimented successfully. The experiment achieved an excellent current density generation result than the anode, which was modified with chemically reduced graphene. The electron transfer was also highly enhanced with this fabrication. Electrochemical exfoliation treatment could also be used to prepare graphene sheet structures. Graphene-modified anode by *in situ* microbial-induced reduction of graphene oxide (GO) was tested, and the current density achieved was 2.34 times than the origin one.

Table 7.2 Anode modification with carbon nanostructures and their performance improvements [42].

Carbon nanostructures	Anode	MFC's efficiency	References
Planar electrodes			
CNTs	Carbon cloth	A maximum power density of 65 mW m^{-2}, an improvement of 250% over the over the non-CNT coated anode (26 mW m^{-2})	Tsai *et al.* (2009)
	Carbon Paper	A maximum power density of −260 mW m^{-2}, associated with the decrease of the anodic resistance	Liang *et al.* (2011)
	Glassy carbon electrode	A maximum current density of 9.7 mA cm^{-2}, 82 times greater than unmodified anode	Peng *et al.* (2010)
	Nickel silicide	A maximum current density of 197 mW m^{-3} of and a power density of 392 mW m^{-3}	Mink *et al.* (2012)
	Carbon paper	A maximum power density of 42.0 mW m^{-2} vs. that of unmodified anode of 16.6 mW m^{-2}	Nambiar *et al.* (2009)
Graphene	Stainless steel mesh (SSM)	A maximum power density of 2,668 mW m^{-2}, which increased by 18 times than 142 mW m^{-2} of SSM	Zhang *et al.* (2011)
Graphene Oxide Nanoribbons	Carbon paper	A maximum power density of 34.2 mW m^{-2} and a current density of 30 A m^{-2}	Huang *et al.* (2011a)
Reduced Graphene Oxide (rGO) Sheets	Carbon cloth	A maximum power density of 2.7 mW m^{-3} and a maximum current of 1.05 +0.09 mA for rGO modified anode vs. that for unmodified anode of 1.7 W m^{-3} and 0.57 + 0.03 mA	Xiao *et al.* (2012)

(*Continued*)

Table 7.2 Anode modification with carbon nanostructures and their performance improvements [42]. (*Continued*)

Carbon nanostructures	Anode	MFC's efficiency	References
Crumpled rGO	Carbon cloth	A maximum power density of 3.6 W m^{-3} and a maximum current of 1.30 + 0.04 mA for crumpled graphene modified anode vs. that for unmodified anode of 1.7 mW m^{-3} and 0. 57 + 0.03 mA	Xu *et al.* (2012)
Bamboo-Like Nitrogen-Doped CNT (N-CNT)	Carbon cloth	A maximum power density of 1.04 W m^{-2} and a maximum current density of 3.63+ 0.06 A m^{-2} for N-CNT modified anode vs. that for unmodified anode of 0.47 W m^{-2} and 0.86 + 0.06 A m^{-2}	Ci *et al.* (2012)
Electrodes			
CNT	Macroscale porous textile (polyester fibers)	A maximum power density of 1,098 mW m^{-2} and a maximum current density of 7.2 A m^{-2} for CNT-textile anode vs. that for carbon cloth anode of 655 mW m^{-2} and 2.8 A m^{-2}	Xie *et al.* (2011)
CNT	Macroporous sponge	A maximum power density of 1.24 W m^{-2} using a batch-fed H-shaped MFC and a maximum power density of 182 W m^{-3} with continuously fed plate-shaped MFC; 2.5 times the previously reported maximum areal power density and 12 times the previously reported maximum volumetric power density	Xie *et al.* (2012a)

(*Continued*)

Table 7.2 Anode modification with carbon nanostructures and their performance improvements [42]. (*Continued*)

Carbon nanostructures	Anode	MFC's efficiency	References
Graphene	Polyurethane sponges with a stainless steel current collector	A maximum power density of 1.57 W m^{-2} and maximum current density of 1.32 A m^{-2} for modified anode vs. CNT coated electrodes from Ci *et al.* (2012)	Xie *et al.* (2012b)
Reduced Graphene Oxide (rGO)	Nickel foam	An optimal volumetric power density of 661 W m^{-3} calculated based on the volume of anode material or 27 W m^{-3} based on the volume of the anode chamber substantially higher than that of plain nickel foam	Wang *et al.* (2013b)
Carbon NanoWeb	RVC	Currently density of 6.8 + 0.3 mA cm^{-2} for NanoWeb-RVC electrodes almost three times higher than RVC	Flexer *et al.* (2013)

7.5.2 Other Methods for Anode Surface Modification

There are various methods utilized for anode surface modification to improve the efficiency of MFCs (Table 7.3). Nitrogen-doped carbon matrix was used as modified anode material to enhance the performance of MFC; this method was quite popular. This method facilitated easier biofilm adhesion on the modified anode surface, providing a higher electron conductivity and current density production.

Cheng *et al.* [21] applied ammonia treatment, where carbon cloth was treated at 700°C with NH_3 gain helium carrier gas. They obtained maximum power density production up to 1,970 mW m^{-2}. Ammonia treatment increased the surface charge that facilitated enhanced bacterial adhesion and electron transport from bacteria to anode. But this method was expensive, which could not be favored for large-scale electricity generation. However, the ammonia-treated carbon mesh showed an excellent

Table 7.3 Surface methods for anode modification [42].

Surface treatment	Anode	MFC's efficiency	Reference
Ammonia Treatment	• Carbon cloth	Ammonia-treated carbon mesh produced a maximum power density of 720 Mw m^{-2} vs. that of mesh chemically cleaned with acetone of 749 mW m^{-2}	Wang *et al.* (2009)
	• Carbon cloth	A maximum power density of 1970 mW m^{-2} using an ammonium-treated carbon cloth anode vs. that of untreated anode of 1,640 mW m^{-2}	Cheng and Logan (2007)
	• Carbon mesh	Ammonia-treated carbon mesh produced a maximum power density of 801 mW m^{-2} vs. that of mesh chemically cleaned with acetone of 624 mW m^{-2}	Wang *et al.* (2009)
Thermal Treatment	• Carbon mesh	Thermal-treated carbon mesh produced a maximum power density of 720 mW m^{-2} vs. that of mesh chemically cleaned with acetone of 624 mW m^{-2}	Wang *et al.* (2009)
	• Carbon brush	A maximum power density of 1,280 mW m^{-2} with treated anode vs. that of untreated anode of 1,020 mW m^{-2}	Feng *et al.* (2010)
Acid Treatment (H_2SO_4)	• Carbon brush	A maximum power density of 1,100 mW m^{-2} with treated anode vs. that of untreated anode of 1,020 mW m^{-2}	Feng *et al.* (2010)

(Continued)

Table 7.3 Surface methods for anode modification [42]. (*Continued*)

Surface treatment	**Anode**	**MFC's efficiency**	**Reference**
Acid Treatment + Thermal Treatment	• Carbon brush	A maximum power density of 1,370 mW m^{-2} with treated anode vs. that of untreated anode of 1,020 mW m^{-2}	Feng *et al.* (2010)
Acid Treatment (HNO_3)	• Graphite granules • Activated carbon fiber felt	The maximum power density reached 8.1 W m^{-3} A maximum power density of 2,066 mW m^{-2} produced by treated anode vs. that of untreated anode of 1,304 mW m^{-2}	Erable *et al.* (2009) Zhu *et al.* (2012)
Ethylenediamine	• Activated carbon fiber felt	A maximum power density of 1641 mW m^{-2} produced by treated anode vs. that of untreated anode of 1,304 mW m^{-2}	Zhu *et al.* (2012)
Electrochemical Oxidation	• Graphite felt • Carbon mesh	A stable current of about 1.13 mA was generated in MFC containing treated anode vs. 0.81 mA of untreated anode. Electrochemical oxidation of carbon mesh in nitric acid of 792 mW m^{-2}, in ammonium nitrate of 736 mW m^{-2}, and in ammonium persulfate of 567 mW m^{-2} vs. that of untreated anode of 552 mW m^{-2}	Tang *et al.* (2011) Zhou *et al.* (2012)

current density generation (801 mW m^{-2}), which was 7% higher than the ammonia-treated carbon cloth (749 m W m^{-2}) cost-effectively.

The conductive polypyrrole structure containing Osmium could be used to immobilize electrons on the anode surface, which could favor enhanced electron transfer efficiency to the anode surface, thus improving the performance of MFCs.

PANI could also be used to modify anode for better efficiency. On the surface of a graphite anode, polyacrylonitrile (PAN) was polymerized, which provided porosity of about 75 cm. Then, electrochemical deposition of CNT over PANI-modified graphite surface was also experimented.

7.6 Cathode Material

As an anode, the cathode is also an integral part of the MFC that also decides the fate of MFCs. For a better performance, cathode should possess high redox potential and can easily capture protons. In MFC, the oxygen gets reduced in a three-phase interface of air (gas), electrode (solid), and electrolyte (liquid) to form water (Figure 7.3). The use of catalyst reduces the activation energy barrier and thus increases the reduction reaction rate. The oxygen molecule gets reduced or ORR takes place in two ways: one is a four-electron pathway and the other one is a two-electron pathway. A four-electron pathway is more reliable than the two-electron pathway. The reactions in an acidic medium can be expressed as below:

$$O_2 + 4H^+ + 4e^- \rightarrow 2H_2O$$

$$O_{2+} 2H^+ + 2e^- \rightarrow H_2O_2$$

$$H_2O_2 + 2H^+ + 2e^- \rightarrow 2H_2O$$

The product formed by the four-electron pathway is water, and the product of the two-electron pathway is hydrogen peroxide.

Most of the materials that are used to the made anode can be used to make cathode too, but cathode should possess the following properties.

- High mechanical strength
- Catalytic properties
- High electronic and ionic conductivity

To improve the performance of MFC, the development of high-performing cathode is an immense priority. Nowadays, most used cathode materials are graphite, carbon cloth, and carbon paper.

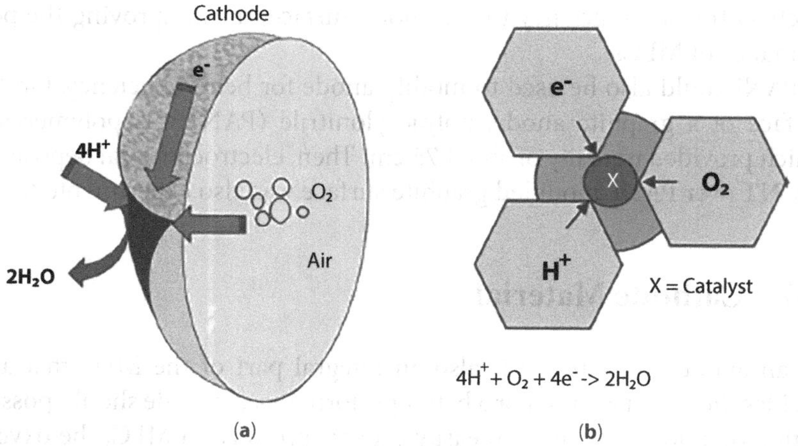

Figure 7.3 The reaction at the cathode takes place at the triple-phase interface of air, solution, and catalyst. (a) Oxygen reduction reaction at three-phase interface producing water as a product. (b) ORR at the three-phase interface of cathode [103].

The major limitation of MFC is low ORR at neutral pH and low temperature, which results in a high overpotential. Since carbon-made cathodes have poor catalytic properties. Thus, an additional catalyst is used to accelerate the ORR. To boost up ORR, many cathode materials are being developed and checked for ORR potential.

To enhance the efficiency of MFC, modification of cathode with highly active catalysts like Pt is used, which is quite popular. Pt treatment with a cathode enhances the reaction rate by reducing the activation barrier or by reducing the cathodic rate. Generally, cathodes can be classified as abiotic and biotic cathodes [105].

7.6.1 Abiotic Cathode

The production of water from oxygen is thermodynamically highly favorable in the presence of a catalyst. To improve the MFC performance, cathodes were modified with a highly active catalyst like platinum (Pt), which suppose to decrease the activation energy barrier and to increase the reduction reaction. Various Pt electrodes were employed as an oxygen-reducing cathode.

Moon *et al.* designed an MFC that uses graphite felt having Pt as cathode that produces power density up to 150 mW m^{-2}, which was higher than the pure graphite cathode. However, Pt is an expensive metal and, due to its catalyst poisoning activity, limits the practical application of

Pt-based electrodes. Pt has been replaced by various nonprecious electrodes like Mn_2O_3 and Fe_2O_3.

Morris *et al.* utilized glucose as a substrate, compared PbO_2 and Pt as a cathode for MFC. The result showed that PbO_2 as a cathode produces two to four times more power than Pt cathode. This concluded that PbO_2 might replace Pt as a cathode catalyst.

When ferricyanide is used as the cathode, the use of catalyst is not required. However, the use of ferricyanide is not suitable for long-term working. The stability of the electrode material for long-term operation is a major issue for large-scale application or industrialization.

7.6.2 Biocathode (Biotic Cathode)

Due to the high cost of Pt, catalyst poisoning and complexity in fabrication lead to the formation of biocathode where microorganisms employ themselves as a catalyst and form a biofilm over the cathode electrode and catalyze the reduction reaction. The most important features of biocathode are self-sustainability and high stability in extremely harsh wastewater conditions.

Biocathode can also be considered as 3D electrode material; it provides an increase in electrode surface area that is an effective way to improve the MFC performance. This facilitates increased microbial attachment and the formation of biofilm over the cathode. Jang *et al.* [22] observed the current generation of MFC with an abiotic cathode. He observed increased current generation with time and stabilized even after 8 weeks of continuous MFC operation, which was higher than the abiotic cathode. From scanning electron microscope (SEM) images, it is concluded that bacteria attach to the cathode and produce nanowires that may accept electrons from the cathode (Figure 7.4).

Kalathil *et al.* [23] developed a (GAC) biological cathode for MFC. It is used for packing minerals in the treatment of wastewater, which is a less expensive and durable material with increased surface area and can be used as a suitable material for a cathode (Figure 7.5).

In the absence of oxygen, nitrate and sulfate act as an electron acceptor at the cathode with denitrifying and sulfate-reducing bacteria, respectively. At biocathode, the microorganism can even produce some chemicals like methane, ethanol, and formic acid. Thus, it can be used as a biosensor to determine biological oxygen demand (BOD) in water. It is also proved by cyclic voltammetry that biocathode is more preferable over abiotic cathode (graphite felt followed by carbon felt).

Biocathode can be classified into two: aerobic biocathode and anaerobic biocathode.

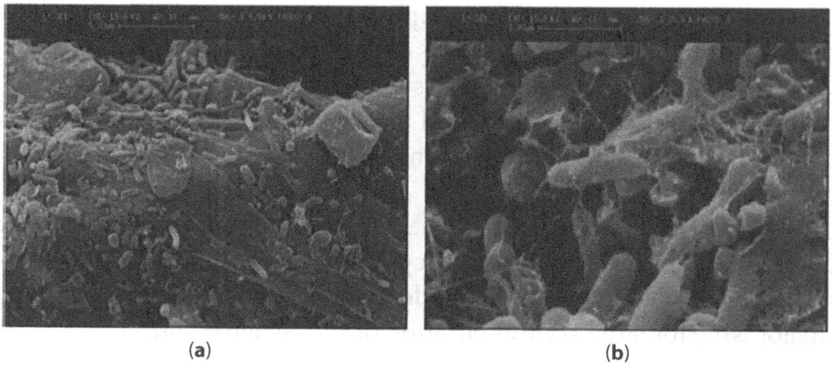

Figure 7.4 SEM image of bacterial adhesion on cathode. (a) 5 micrometer (b) 1 micrometer

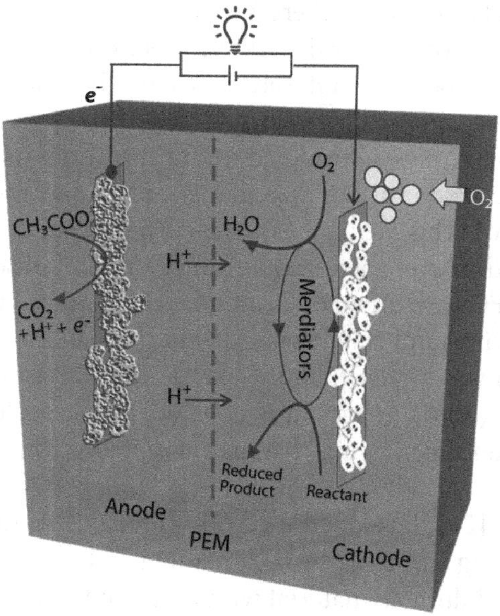

Figure 7.5 A setup of MFC using biological cathode, where oxygen and the other chemical reactants are reduced. This system is can be used to produce electricity or to perform biochemical synthesis reactions to synthesize value-added chemicals [103].

At aerobic biocathode, oxygen acts as an electron acceptor, and hydrogen peroxide acts as an intermediate of the reaction. Various other transition metals are the electron mediator between oxygen and cathode electron-generated by the biofilm reduce Fe (III) to Fe (II), which is subsequently oxidized by oxygen. In this way, electrons are transferred from the cathode to oxygen (terminal electron acceptor).

At anaerobic biocathode, in the absence of oxygen, nitrate and sulfate act as terminal electron acceptors. The cathodic potential of chemically active species like manganese, nitrate, and iron is comparable with oxygen. Over this, the major advantage of anaerobic biocathode is the prevention of loss of electrons through O_2 that will later diffuse to the anode via PEM.

7.6.3 Cathode with Pt-Based Catalyst

Cathode without catalyst coating is not enough to perform the whole oxidation-reduction reaction forming water using oxygen. Due to its high activation energy barrier, it requires catalyst coating on the cathode. Pt is the most commonly used catalyst for ORR as it acts as 3D electrode material providing higher surface area and less overpotential for ORR. Zhang *et al.* showed that the maximum power density generation of 1,610 W/m^2 can be achieved using a double layer of PDMS/carbon loaded with Pt. Ghasemi *et al.* [24] reduced Pt. loading by depositing over CNTs, showing improvement in the ORR activity for CNT-Pt composite.

7.6.4 Cathode with Non–Pt-Based Catalyst

Pt is an expensive metal and, due to its pH sensitivity, poisoning activity, and long-term instability, limits its use for large-scale production. Thus, there is a need to develop an alternative without the use of Pt catalyst usually called a non–Pt-based catalyst. Other transition metals like iron phthalocyanine are considered as a catalyst, due to its Π-Π interaction among the metal and carbon consisting of the aromatic ring, leading to the rapid electron transfer. Iron phthalocyanine supported with amino-functionalized multiwall CNT_S can get considerably high power density with CNT cathodes. It was observed that the power density generation was much greater than the platinum-carbon cathode.

Chenget *et al.* used metal macrocyclic compounds such as CoTMPP as ORR catalyst with comparable Columbia efficiency to that of Platinum [14]. Lead dioxide can replace Pt as a catalyst having low cost and shows four times greater power densities. Platinum as a catalyst can be easily replaced by metal-based oxide catalyst like MnO_2 (manganese dioxide), which use as a shutting mechanism for ORR and found that MnO_2 catalyzed cathodes have a much larger cathode current than a graphite electrode.

Zhang *et al.* compared three forms of manganese dioxide α, β, and γ and concluded that MnO_2 used PVDF (polyvinylidene fluoride) as a binder to get deposited over glass carbon and observed catalytic properties of β MnO_2. β MnO_2 has properties quite close to Pt catalyst [25]. Apart from MnO_2 and palladium, Pt-like transition metals are tested because of their low cost and excellent catalytic properties. Overpotential for catalytic hydrogen production, palladium nanoparticle has a very low ORR as explained by Hang *et al.* and showed the high stability in the linear sweep voltammetry result [26].

7.7 Conclusion

MFC is a future perspective technology, which would be implemented on large scale as a natural renewable eco-friendly aspect for commercial use. MFC technology can be utilized for various aspects such as wastewater treatment, mainly energy or electricity production, and various products such as jet engine fuel production. To utilize this eco-friendly technological approach for commercial use, it needs to be cost-effective and highly efficient to produce a large amount of energy or desired product. Many researchers are trying their best to improve its performance and are experimenting on various electrode materials, which should require low cost to manufacture. Based on the experiments conducted in the last decade, it is concluded that, to improve MFC performance, there is a requirement to improve the quality of electrode material. The surface area of anode material and its electron conductivity ability highly affects the MFC performance. The large surface area of the anode facilitates more amount of biomass adhesion. Researchers showed that improvement in anode material directly enhances MFC output.

Different materials such as carbon-based anode materials such as carbon cloth, carbon mesh, graphite, GC, and RVC; CNT-based anode materials; and composite materials such as graphene, ceramic, and stainless steel have been experimented by researchers to manufacture an efficient anode. Apart from this, several anode surface treatments have also been tested to improve the performance of MFC and achieved good results. But for commercial use, it is not good enough. It needs to be highly improved to apply on large scale. Electrode materials play a vital role economically and in the performance of MFC. Electrodes used in MFC need to be very low in cost and highly efficient for large-scale long-term commercial production.

Cathode needs an increase in the catalytic property of electrodes due to poor ORR. Pt is considered as one of the best catalysts, but it limits accessibility for commercial application due to its higher cost. This makes MFC unfeasible to apply for large-scale production. Various transition metals oxides like MnO_2 came as a promising material for ORR with low cost. Microorganisms can also act as a suitable catalyst that interacts with the electrode surface and optimize the performance of the MFCs.

References

1. Wang, H. and Ren, Z.J., A comprehensive review of microbial electrochemical systems as a platform technology. *Biotechnol. Adv.*, 31, 1796–807, 2013, https://doi.org/10.1016/j.biotechadv.2013.10.001.
2. Potter, M.C., Electrical effects accompanying the decomposition of organic compounds. *Proc. R. Soc. Lond. Ser. B Contain. Pap. Biol.*, 84, 260–76, 1911, https://doi.org/10.1098/rspb.1911.0073, ScD. MA.
3. Wilkinson, S., "Gastrobots"—Benefits and Challenges of Microbial Fuel Cells in FoodPowered Robot Applications. *Auton. Rob.*, 9, 99–111, 2000, https://doi.org/10.1023/A:1008984516499.
4. Freguia, S., Rabaey, K., Yuan, Z., Keller, J., Electron and carbon balances in microbial fuel cells reveal temporary bacterial storage behavior during electricity generation. *Environ. Sci. Technol.*, 41, 2915–21, 2007, https://doi.org/10.1021/es062611i.
5. Zhou, M., Chi, M., Luo, J., He, H., Jin, T., An overview of electrode materials in microbial fuel cells. *J. Power Sources*, 196, 4427–35, 2011, https://doi.org/10.1016/j.jpowsour.2011.01.012.
6. Watanabe, K., Recent developments in microbial fuel cell technologies for sustainable bioenergy. *J. Biosci. Bioeng.*, 106, 528–36, 2008, https://doi.org/10.1263/jbb.106.528.
7. Zou, Y., Xiang, C., Yang, L., Sun, L.-X., Xu, F., Cao, Z., A mediatorless microbial fuel cell using polypyrrole coated carbon nanotubes composite as anode material. *Int. J. Hydrogen Energy*, 33, 4856–62, 2008, https://doi.org/10.1016/j.ijhydene.2008.06.061.
8. Liu, H., Ramnarayanan, R., Logan, B.E., Production of electricity during wastewater treatment using a single chamber microbial fuel cell. *Environ. Sci. Technol.*, 38, 2281–5, 2004, https://doi.org/10.1021/es034923g.
9. Chaudhuri, S.K. and Lovley, D.R., Electricity generation by direct oxidation of glucose in mediatorless microbial fuel cells. *Nat. Biotechnol.*, 21, 1229–32, 2003, https://doi.org/10.1038/nbt867.
10. Ahn, Y. and Logan, B.E., Effectiveness of domestic wastewater treatment using microbial fuel cells at ambient and mesophilic temperatures. *Bioresour. Technol.*, 101, 469–75, 2010, https://doi.org/10.1016/j.biortech.2009.07.039.
11. Logan, B., Cheng, S., Watson, V., Estadt, G., Graphite fiber brush anodes for increased power production in air-cathode microbial fuel cells. *Environ. Sci. Technol.*, 41, 3341–6, 2007, https://doi.org/10.1021/es062644y.

12. He, Z., Minteer, S.D., Angenent, L.T., Electricity generation from artificial wastewater using an upflow microbial fuel cell. *Environ. Sci. Technol.*, 39, 5262–7, 2005, https://doi.org/10.1021/es0502876.
13. Kim, J.R., Jung, S.H., Regan, J.M., Logan, B.E., Electricity generation and microbial community analysis of alcohol powered microbial fuel cells. *Bioresour. Technol.*, 98, 2568–77, 2007, https://doi.org/10.1016/j.biortech.2006.09.036.
14. Liang, P., Fan, M., Cao, X., Huang, X., Huang, Z., Wang, C., [Electricity generation using the packing-type microbial fuel cells]. *Huan Jing Ke Xue*, 29, 512–7, 2008.
15. Ketep, S.F., Bergel, A., Calmet, A., Erable, B., Stainless steel foam increases the current produced by microbial bioanodes in bioelectrochemical systems. *Energy Environ. Sci.*, 7, 1633–7, 2014, https://doi.org/10.1039/C3EE44114H.
16. Qiao, Y., Li, C.M., Bao, S.J., Bao, Q.L., Carbon nanotube/polyaniline composite as anode material for microbial fuel cells. *J. Power Sources*, 170, 79–84, 2007, https://doi.org/10.1016/j.jpowsour.2007.03.048.
17. Tsai, H.-Y., Wu, C.-C., Lee, C.-Y., Shih, E.P., Microbial fuel cell performance of multiwall carbon nanotubes on carbon cloth as electrodes. *J. Power Sources*, 194, 199–205, 2009, https://doi.org/10.1016/j.jpowsour.2009.05.018.
18. Sharma, T., Mohana Reddy, A.L., Chandra, T.S., Ramaprabhu, S., Development of carbon nanotubes and nanofluids based microbial fuel cell. *Int. J. Hydrogen Energy*, 33, 6749–54, 2008, https://doi.org/10.1016/j.ijhydene.2008.05.112.
19. Xiao, L., Damien, J., Luo, J., Jang, H.D., Huang, J., He, Z., Crumpled graphene particles for microbial fuel cell electrodes. *J. Power Sources*, 208, 187–92, 2012, https://doi.org/10.1016/j.jpowsour.2012.02.036.
20. Zhang, Y., Mo, G., Li, X., Zhang, W., Zhang, J., Ye, J. *et al.*, A graphene modified anode to improve the performance of microbial fuel cells. *J. Power Sources*, 196, 5402–7, 2011, https://doi.org/10.1016/j.jpowsour.2011.02.067.
21. Cheng, S. and Logan, B.E., Ammonia treatment of carbon cloth anodes to enhance power generation of microbial fuel cells. *Electrochem. Commun.*, 9, 492–6, 2007, https://doi.org/10.1016/j.elecom.2006.10.023.
22. Jang, J.K., Kan, J., Gorby, Y., Hsu, L., Kim, B., Nealson, K., Electricity Generation by Microbial Fuel Cell Using Microorganisms as Catalyst in Cathode. *J. Microbiol. Biotechnol.*, 23, 1765–1773, 2013, https://doi.org/10.4014/jmb.1310.10117.
23. Kalathil, S., Lee, J., Cho, M.H., Granular activated carbon based microbial fuel cell for simultaneous decolorization of real dye wastewater and electricity generation. *N Biotechnol.*, 29, 32–7, 2011, https://doi.org/10.1016/j.nbt.2011.04.014.
24. Ghasemi, M., Shahgaldi, S., Ismail, M., Kim, B.H., Yaakob, Z., Wan Daud, W.R., Activated carbon nanofibers as an alternative cathode catalyst to platinum in a two-chamber microbial fuel cell. *Int. J. Hydrogen Energy*, 36, 13746–52, 2011, https://doi.org/10.1016/j.ijhydene.2011.07.118.

25. Zhang, L., Liu, C., Zhuang, L., Li, W., Zhou, S., Zhang, J., Manganese dioxide as an alternative cathodic catalyst to platinum in microbial fuel cells. *Biosens. Bioelectron.*, 24, 2825–9, 2009, https://doi.org/10.1016/j.bios.2009.02.010.
26. Huang, Y.-X., Liu, X.-W., Sun, X.-F., Sheng, G.-P., Zhang, Y.-Y., Yan, G.-M. *et al.*, A new cathodic electrode deposit with palladium nanoparticles for cost-effective hydrogen production in a microbial electrolysis cell. *Int. J. Hydrogen Energy*, 36, 2773–6, 2011, https://doi.org/10.1016/j.ijhydene.2010.11.114.

8

Extensive Use of Photocatalysis and Photoelectrochemical Methods for Wastewater Reclamation

Namrata Khanna[1*], Tanushri Chatterji[2†], Virendra Yadav[3], Tanya Bhagat[2], Shalini Sharma[2] and Sadanand Pandey[4‡]

[1]*Department of Biochemistry, M A Rangoonwala College of Dental Sciences and Research Centre, Pune, Maharashtra, India*
[2]*Department of Biosciences, IMS Ghaziabad (University Courses Campus), Adhyatmik Nagar, Uttar Pradesh, Ghaziabad, India*
[3]*Department of Life Sciences, Hemchandracharya North Gujrat University, Patan, Gujrat, India*
[4]*School of Bioengineering and Food Technology, Faculty of Applied Sciences and Biotechnology, Shoolini University, Solan, Himachal Pradesh, India*

Abstract
The conventional procedures employed for removal of resolute organic toxicants in water have proved to be insufficient and inept, necessitating development of innovative strategies for wastewater treatment. Photoelectrochemical reactions are being accepted as favorable alternative strategies along with other oxidative procedures for reclaiming wastewater. The combined action of photochemical and electrolytic reactions is highly effective against various toxicants. This synergistic action enhances generation of hydroxyl radicals and increases mineralization. The current chapter elaborates basic mechanisms of various photoelectrochemical processes inclusive of photoelectrocatalysis, photocatalytic fuel cells, solar or photoelectro-Fenton, and photoelectroperoxone. The use of photoelectrochemical processes in treatment of oil mill effluents, landfill leachate, emerging contaminants, tannery effluents, and textile waste is discussed. The relevance of photocatalytic fuel cells for minimizing use of electrical procedures is assessed.

**Corresponding author*: namratakhanna6@gmail.com
†*Corresponding author*: tanushri@gmail.com
‡*Corresponding author*: sadanand.au@gmail.com; spandey@shooliniuniversity.com

Sadanand Pandey, Elvis Fosso-Kankeu and Soumya Pandit (eds.) Synergy of Bio-Chemical Processes for Photocatalytic and Photoelectrochemical Wastewater Treatment, (189–244) © 2024 Scrivener Publishing LLC

Photocatalysts play a significant role as effective and viable oxidative agents for use in wastewater treatment. We focus on innovative photocatalysts being developed for increasing efficiency in wastewater reclamation. Use of semiconductor photocatalysts has resulted in multidisciplinary research, owing to their unique ability of converting solar energy to chemical energy. We review various studies on development of innovative photocatalysts, being produced by novel strategies with pertinent modifications, resulting in enhanced efficiency for degrading several toxicants in water, like antibiotics, dyes, phenols, and several other organic compounds.

Keywords: Wastewater reclamation, photoelectrochemical process, photoelectrocatalysis, semiconductors, photocatalytic fuel cells

Abbreviations

2,4-D	2,4-dichlorophenoxyacetic acid
2,4-DCP	2,4-dichlorophenol
4-CPA	4-chlorophenoxyacetic acid
AO	anodic oxidation
AOP	advanced oxidation process
BDD	boron-doped diamond
BOD	biochemical oxygen demand
CB	conduction band
CFU	colony-forming unit
COD	chemical oxygen demand
EAOP	electrochemical AOP
EAOPs	electrochemical advanced oxidation processes
Ebg	bandgap energies
EC	electrocoagulation
EC	emerging contaminant
EDC	endocrine-disrupting chemicals
EF	electro-Fenton
g-C_3N_4	graphitic carbon nitride
GO	graphene oxide
magnetic HMS	magnetic hexagonal mesoporous silica
MCPA	2-methyl-4-chlorophenoxy acetic acid
NAs	nanoarrays
NBs	nanobelts
NF	nano filtration
NRs	nanorods
NTs	nanotubes

PEF	photoelectro-Fenton
PEP	photoelectro-peroxone
PF	photo-Fenton
PFCs	photocatalytic fuel cells
POP	persistent organic pollutants
PPCP	personal care products
RO	reverse osmosis
ROS	reactive oxygen species
SOC	synthetic organic compounds
SPEF	solar PEF
SPF	solar PF
SPR	surface plasmon resonance
TMP	trimethoprim
TOC	total organic carbon
TOC	total organic carbons
VB	valence band
VLD	visible light driven
VUV	vacuum UV

8.1 Introduction

Inaccessibility to potable water, scarce water supply, and deficient storage affect billions worldwide [1]. With increasing population, the industrial, agricultural, and domestic demand for water increases dramatically, causing an enhanced surge in freshwater usage coupled with the effects of climate change that many countries will face water stress in coming years [2]. Hence, the use of water and reclamation of wastewater will be significant in future. In addition, there is another challenge, which involves extraction of substances like organic micropollutants, for which current urban wastewater treatment reactors are unequipped [3]. Thus, novel water treatment strategies need to be developed for treating municipal or industrial wastewater prior to their release into water bodies. An appropriately planned photoelectrocatalytic reactor is a probable solution, and, if required, it can be integrated with other techniques to generate a multiple-step water treatment plan. The expression "*photoelectrocatalysis*" as defined in the IUPAC Recommendations 2011 is "*electrochemically assisted photocatalysis. The role of the photocatalyst is played by a photoelectrode, often a semiconductor*" [4].

Water is requisite for every living entity; hence, removal of contaminants from water is essential. Wastewater is generated at an estimated rate of 22,900 million liters per day (MLD), which includes wastewater from domestic use and 13,500 MLD of contaminated from industries in an urban domain. Approximately 26% wastewater undergoes treatment prior to their discharge, while the remaining is left untreated. Local wastewater has 70% of organic and 30% of inorganic waste material. Treatment of industrial waste is complicated, owing to the hazardous content, which is difficult to remove naturally [5, 6]. With the advent of industrialization in the last century, manufacturing required large amounts of water. Freshwater is required in tons each year for production of synthetic organic compounds (SOCs). Textile industry consumes 80% of total dyes produced and generates large amounts of wastewater. Dyes in water affect photosynthesis in aquatic plants and provide a favorable environment for detrimental organisms that deteriorate the ecosystem [7]. The waste is composed of various hazardous components like heavy metals, dyes, metals, petrochemical compounds, thickeners, and stabilizers [8]. Pollution caused by SOC production leads to various health and sustainability problems [9, 10].

Water is ubiquitous; thus, approximately 80% of diseases are waterborne, and water pollution leads to 3.1% of deaths. The organic and inorganic compounds in wastewater may be taken up by microorganisms, affecting the rate of photosynthesis in the plankton [11]. These compounds are highly stable and resistant to degradation; hence, their removal before discharge is essential. Many conventional physicochemical methods have been developed, like adsorption, oxidation, precipitation, coagulation, use of nanoparticles, and electrochemical procedures [12–14].

Photoelectrochemical strategy is an advanced oxidation process (AOP), mostly employed for degrading toxic material in wastewater. The light and electrical energies amalgamate to efficiently remove contaminants from wastewater. The current chapter elaborates the basic mechanism of the photoelectrochemical process, use of photocatalysts, photocatalytic fuel cells (PFCs), and extensive-scale innovative wastewater conduct strategies.

8.2 Mechanism Describing the Photoelectrochemical Processes

Photoelectrochemical process is a modification of the AOP. The photons and electrical energy are involved in synchronization during the treatment procedure. Electrochemical processes involve many AOPs, applied either independently or together. Most AOPs generate free OH radicals for degradation of SOCs. Typically, SOCs are degraded by two strategies:

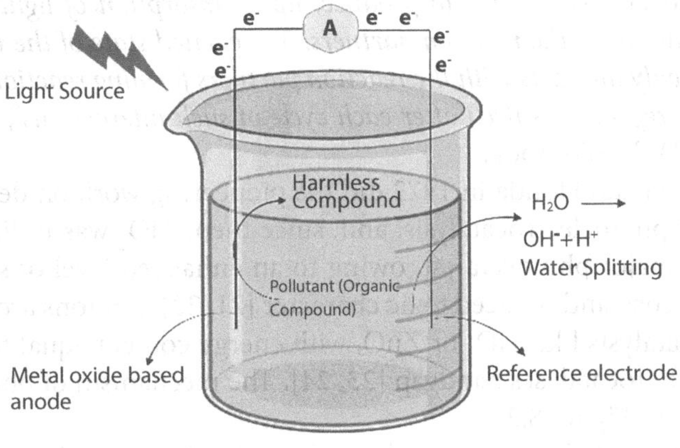

Figure 8.1 Mechanism of photoelectrochemical process for treating wastewater [16].

one is through transformation of pollutants into simpler biodegradable compounds, which is accomplished by transferring electrons directly or indirectly and is called the electrochemical conversion, whereas the second strategy involves mineralization of organic compounds through free radicals and is called the electrochemical combustion. In the photoelectrochemical procedure, photo-assisted mediators that depend on the wavelength range of UV and visible spectrum are used. Photo-assisted mediators are activated by photons, which generate free OH radicals for oxidation of organic molecules. Free radicals, generated by activation of metallic nanomaterials, are utilized by oxidative processes, which convert organic compounds to simpler products. Semiconductor-based catalysts are significant for the treatment of pollutants in water. Metals are good conductors and effectively generate free hydroxyl radicals for efficient remediation of several organic pollutants [15] (Figure 8.1).

8.3 Photoelectrocatalysis

In accordance with IUPAC Gold Book, the process of photocatalysis is defined as "*change in the rate of a chemical reaction or its initiation under the action of ultraviolet, visible, or infrared radiation in the presence of a substance—the photocatalyst—that absorbs light and is involved in the chemical transformation of the reaction partners*". Photocatalysis is extensively employed for inactivation of microorganisms [17], reducing metal ions to decrease toxicity [18] and degrading chemicals like dyes, phthalates, pharmaceuticals, and pesticides [19, 20].

A photocatalyst is "*able to produce, upon absorption of light, chemical transformations of the reaction partners. The excited state of the photocatalyst repeatedly interacts with the reaction partners forming reaction intermediates and regenerates itself after each cycle of such interactions*", according to the IUPAC Gold Book.

Fujishima and Honda in 1972 did the pioneering work on devising the process of photoelectrocatalysis, and, since then, TiO_2 was utilized in the form of a major photocatalyst, owing to an enhanced level of steadiness, decreased cost, and reduced toxic character [21, 22]. Photons are absorbed by photocatalysts like TiO_2 or ZnO, with energy content equal to or more than the photocatalyst's bandgap [23, 24]. The mechanism of photocatalysis is given in Figure 8.2.

The efficacy of photocatalysis depends on the charge parting and recombination of electron–holes that exist in pairs. Hence, a reduction in the recombination of electron–hole pairs, is required for enhancing the photo-oxidation of the contaminant. Typically, photocatalysts exhibit a wide bandgap and can captivate energy only on UV exposure, which constitutes a miniscule 5% of the energy of sun [25]. Visible light is favored for photocatalytic application in large-scale use, because it constitutes more than 40% of the solar spectrum [26]. Considerable studies have been conducted to develop different visible-light–dependent photocatalysts, although limitations like less absorption of visible light and low quantum efficiency have not been resolved. Lately, several studies have been conducted like doping of semiconductors with metals or non-metals and development of a semiconductor heterojunction through amalgamation with a second semiconductor, which enhances efficiency of the photocatalyst and overcomes limitations of conventional semiconductors [27, 28].

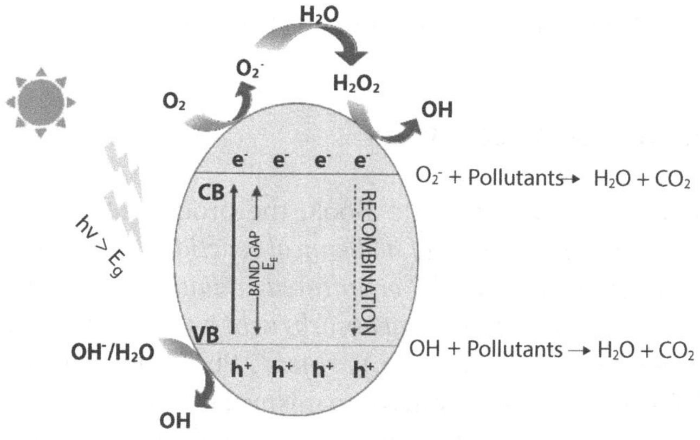

Figure 8.2 Schematic diagram explaining photocatalysis.

Semiconductor doping with non-metallic elements enhances the action to visible light and ameliorates effectiveness of the photocatalytic semiconductor [29]. Development of a composite semiconductor is an effective technique, which increases the visible light absorbance and improves their efficacy as a photocatalyst [30]. The development of a composite semiconductor utilizes sensitization of dyes, wherein the semiconductor functions like a sensitizer, rather than a dye, owing to a reduced bandgap energy. Further, the two integrated semiconductors have discrete energy levels, playing a key role in segregation of charge carriers [31]. Generally, a semiconductor possessing wide bandgap conjugates with another semiconductor of lower bandgap. Consequently, energy drawn up by one semiconductor gets conveyed to another across the heterojunction, further enhancing the efficiency of the composite semiconductor as a photocatalyst [26, 32]. Over the years, photocatalysis has been employed in decontamination of water from several inorganic and organic pollutants that have percolated water bodies. The most common contaminants are the synthetic dyes, which have been degraded through photocatalysis [33, 34]. Chen *et al.* investigated the degradation of methyl orange with ZnO, which acts as a photocatalyst. They concluded that ZnO prepared from oxalic acid and zinc acetate in a molar ratio of 4:1 with subsequent calcination at 400°C, revealed a maximum 99.7% efficiency within 30 min of UV irradiation. The enhanced competence of degrading contaminants was related to an appropriate measure of chemicals and the optimum temperature employed for synthesizing the semiconductor. The minor particle size and improved crystallinity of the photocatalyst yielded better results. Photocatalysis is now being applied in treatment of several emerging contaminants (ECs) like pharmaceuticals, chemicals disrupting endocrine activity [35], and herbicides [36]. Efficiency of BiOCl microsphere photocatalyst was observed for carbamazepine removal on exposure to visible light [37]. The study reported better efficiency of the BiOCl-10 photocatalyst in degrading 70% carbamazepine in 180 min, owing to enhanced carbamazepine adsorption, effectively segregating electron–hole pairs and increasing absorption in the visible region. Another investigation stressed on increasing the surface area and porous character of bulk g-C_3N_4 by performing an acid treatment in extrication of bisphenol A. Post–acid treatment, the exfoliated g-C_3N_4 displayed an heightened degradation effectiveness of 99%, owing to an increase in surface area, which furnished enhanced active sites for visible light absorption [38]. Another study employed a template of natural kapok fiber for synthesis of TiO_2 hollow fibers, which are used for extricating paraquat herbicide upon UV irradiation [39]. The investigations reported that hollow fibers composed of TiO_2, produced at 450°C, had maximum effectiveness, with 50% efficiency in degradation, suggestive of highest photocatalytic activity at the calcination temperature [40]. The basic redox reactions take place at the photocatalytic surface, resulting in degradation of the contaminant through reactive oxygen species (ROS) (Figure 8.3). In step (1), the incident photon is absorbed by the semiconductor, and the photon possesses energy more than the bandgap of semiconductor. If recombination is circumvented, then the electron (e^-) and hole (h^+), generated by photons, move to the external area of semiconductor, where oxidation reactions in steps (3) to (5) and reduction reactions in steps (6) to (9) complete the process [41].

For a complete oxidation reaction, the valence band edge potential (E_{VB}) should bear a higher positive charge than the reaction potential required; like the hydroxyl radical (•OH in step 5) has +2.29 V as the reduction potential, and, with TiO_2, E_{VB} is +2.69 V; hence, it effectively oxidizes water to generate the•OH radical [43]. For reactions involving reduction,

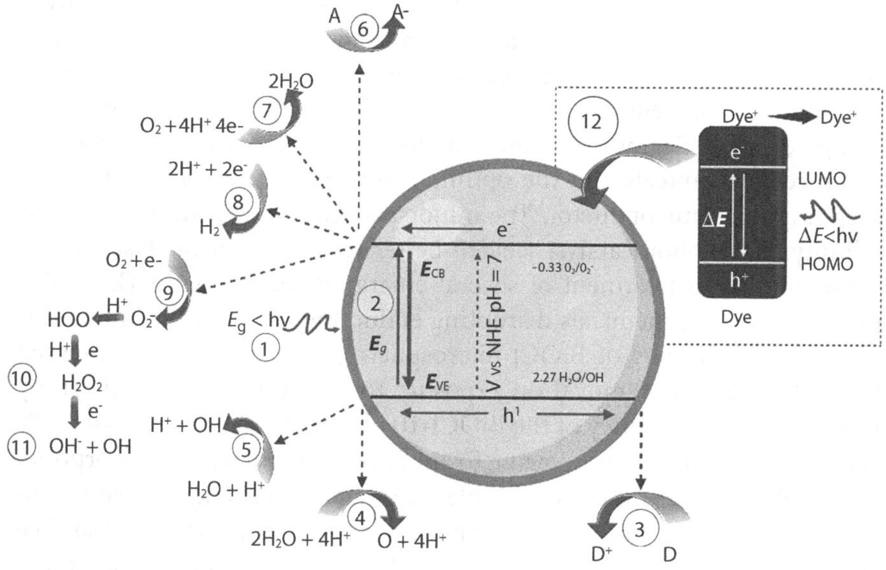

Figure 8.3 Representation of photocatalysis by a semiconductor. (1) Absorption of photons, (2) excitation by photons and recombination, (3) transfer of electron by donor, (4) evolution of oxygen, (5) generation of hydroxyl radicals upon oxidation of water, (6) an acceptor receives electrons, (7) reduction of proton and oxygen during formation of water, (8) protons reduced to hydrogen atom, (9) superoxide formation from oxygen reduction, (10) hydrogen peroxide generation, (11) hydroxyl radical generation, and (12) sensitization of the dye and exchange of electrons toward the conduction band (* excited state) [42].

the conduction band edge potential (E_{CB}) should have a higher negative charge in comparison to the preferred reduction potential of reaction; for instance, the superoxide radical anion ($O^{\cdot}{}_2{}^-$) carries a reduction potential of −0.33 V and TiO_2 (anatase) has an E_{CB} of −0.51 V, making TiO_2 thermodynamically effective in reducing O_2 to $O^{\cdot}{}_2{}^-$ [44].

Photoelectrochemical and photoelectrocatalytic techniques have been extensively investigated for application in wastewater reclamation and water splitting. The photoelectrochemical procedure involving splitting of water was first elucidated by researchers Fujishima and Honda [21], which was later studied on the electrochemical basis by Bockris *et al.* [45], although its actual application in wastewater reclamation was initiated in the beginning of 21st century [46]. The above process employs a scheme involving electrolysis, which constitutes a thin layer photo anode, irradiated with a constant anodic bias potential (E_{anod}), a constant anodic current density (j_{anod}), or a constant cell potential (E_{cell}) [29]. It raises removal of photoexcited $E_{CB}{}^-$ from the outer electric field, accordingly giving an effective parting of the $E_{CB}{}^-/h_{vb}{}^+$ pairs, causing inhibition of reactions in Equations (8.1) to (8.5) [47, 48]

$$\text{photocatalyst} + hv \rightarrow E_{CB}{}^- + h_{vb}{}^+ \tag{8.1}$$

$$h_{vb}{}^+ + H_2O \rightarrow OH + H^+ \tag{8.2}$$

$$h_{vb}{}^+ + OH^- \rightarrow OH \tag{8.3}$$

$$h_{vb}{}^+ + R \rightarrow R^+{}_{OX} \tag{8.4}$$

$$E_{CB}{}^- + O_2 \rightarrow O_2{}^- \tag{8.5}$$

The recombination of photon-excited $E_{CB}{}^-/h_{vb}{}^+$ ratio, when inhibited, may enhance the rate of redox reactions, causing increase in photo-anodic catalysis, along with enhanced production of photon-excited $h_{vb}{}^+$ and increased degradation of organic contaminants. Hence, the lifetime of $h_{vb}{}^+$ increases, and various prospective direct or indirect techniques of organic degradation can be elucidated. Degradation is direct when adsorption occurs on the photo-anodic surface and is indirect when ·OH radical formation increases, as described in Equation (8.2) [48]. The photocatalysts can be reused for further PEC degradation. For example, when an anodic potential is passed through a semiconductor (M), the organic contaminants are degraded by electrochemical oxidation (ECO). The process of ECO oxidizes water into hydroxyl radical, forming M(·OH) in the process [Equation (8.6)]

$$M + H_2O \rightarrow M(\cdot OH) + H^+ + E_{CB}^- \quad (8.6)$$

Photon-excited h_{vb}^+ is highly oxidizing, whereas the excited E_{CB}^- can be a probable reducing moiety [Equation (8.1)], which mineralizes organic compounds. The photon-excited E_{CB}^- reacts with O_2 to release a superoxide radical O_2^-, as given in Equation (8.5) [49].

Photoelectrocatalysis has a potential advantage over ordinary photocatalysis as it enhances the generation of radicals upon amalgamation of photocatalysis and electrolysis [50]. Photocatalysis is a broadly studied AOP, used for mineralization of multiple-phase contaminants. When irradiated with visible, UV, and near-UV spectra, semiconductor photocatalysts such as zinc oxide (ZnO), vanadium oxide (V_2O_5), titanium dioxide (TiO_2), bismuth vanadate ($BiVO_4$), or tungsten trioxide (WO_3) act as potent generators of reactive oxygen species (·OH and $\cdot O_2^-$) [51].

8.3.1 Types of Photocatalysts

An efficacious breakdown of contaminants depends on the semiconductor moieties opted as photocatalysts and on the inherent photocatalytic properties exhibited. Typically, the photocatalysts employed are metal oxides with pertinent gaps in energy within required valence band (VB), along with an empty conduction band (CB) for generating photon-excited E_{CB}^-/h_{vb}^+ pairs on exposure to light. A few semiconductors employed as photocatalysts for wastewater reclamation are explained in the following sections.

8.3.1.1 Metal Oxides

The effective use of photoelectrocatalytic process for removal of organic contaminants depends on the photocatalytic properties of semiconductors and their use as photo anodes. Typically, metal oxides with pertinent bandgap energies (Ebg) within the required VB and empty CB are employed to generate the photon-excited pairs of E_{CB}^-/h_{vb}^+. A variety of photocatalysts are employed for application to efficiently decontaminate water from organic and inorganic toxicants. The most frequently used photocatalyst is TiO_2, which has three crystalline structures, named anatase, rutile, and brookite, which have energy bands (Ebg) of 3.23 eV, 3.02 eV, and 3.14 eV, respectively. Anatase TiO_2 is the most active crystalline structure in both photocatalytic and photoelectrocatalytic processes. The most common commercially available semiconductor is Degussa P-25, which is widely used for reclamation due to a less toxic, luminous, and non-reactive feature. Anatase TiO_2 is predominantly used as a metal oxide photocatalyst used for remediation and effective treatment of pollution [52].

Another metal oxide zinc oxide (ZnO) is found extensively in the crust and is significantly cost effective than TiO_2 and is non-toxic to humans. ZnO is present as hexagonal wurtzite crystals and cubic zinc blende. Hexagonal wurtzite is highly stable electrochemically with an E_{bg} of 3.4 eV and is widely used with solar irradiation [53]. ZnO is available as thin transparent films, which enhances photon absorption within the semiconductor, increasing generation of photon-induced pairs of E_{CB}^{-}/h_{vb}^{+}. Tungsten oxide (WO_3) is also employed as a photocatalyst, with crystalline structures that are relatively temperature dependent. WO_3 has an E_{bg} value between 2.5 eV and 2.7 eV, which is considerably lesser than values exhibited by TiO_2; hence, WO_3 can more efficiently generate E_{CB}^{-}/h_{vb}^{+} on UV irradiation [54]. The initial use of WO_3 as a photo-anode in wastewater treatment was described by researchers Hepel and Luo [55]. A major disadvantage with WO_3 is a higher toxicity than TiO_2 [56]. Another potential metal oxide semiconductor MnO has a lower E_{bg} of 1.3 eV, is more cost effective, and has increased surface area and better electrochemical stability [57], along with non-toxic character [58]. Combining MnO with polyaniline can further enhance the effectiveness of MnO as a photocatalyst, as the combination with polyaniline's E_{bg} (2.8 eV) reduces the E_{CB}^{-}/h_{vb}^{+} recombination. Furthermore, another semiconductor SnO_2 is unsuited for use as a photocatalyst [59], but, as a non-conductor, SnO_2 is conveniently modified upon doping with Sb, which significantly enhances the electrical conductivity of SnO_2. The modified Sb-SnO_2 has lower electrochemical stability and a short lifetime, which is conducive for its use as a photocatalyst [60]. An additional semiconductor β-PbO_2 may be used as a photocatalyst as it has a comparatively lower E_{bg} of 1.4 eV [61], although β-PbO_2 is not employed in acidic conditions, where the toxic Pb metal can be leached in the medium. A few stable semiconductors like $BiVO_4$ with E_{bg} 2.5 eV and Bi_2WO_5 with E_{bg} 2.8 eV [62], on exposure to visible spectra and $BiPO_6$ with E_{bg} 3.8 eV [63], on exposure to UV rays, exhibit remarkable efficiency and improved degradation of organic contaminants through photoelectrocatalytic process.

8.3.1.2 Modified or Doped Metallic Oxides

The effectiveness of photoelectrocatalysis is largely dependent on characteristics and absorption of incident light, velocity of redox reactions on the surface of photocatalyst, and the recombination rate of photon-excited E_{CB}^-/h_{vb}^+ pairs. For increasing the effectiveness of photo(electro)catalysts, several modifications have been implemented, such as (1) remodeling of the metal oxides into nanotubes (NTs), nanobelts (NBs), nanorods (NRs), or nanoarrays (NAs); (2) modifying the semiconductor photocatalysts with metals like Fe, Cr, and Cu or non-metals like N and B; (3) combining semiconductors with metals like Pd, Au, and Ag, oxides like SiO_2, WO_3, and Sb_2S_3, along with carbon containing substances like graphene and carbon fabric; and (4) generating innovative materials with titanium like $TiNbO_5$ and cubic double-perovskite like $CaCu_3Ti_4O_{12}$. Conventional photocatalysts like TiO_2, when doped with metals/non-metals, will exhibit an innovative VB and CB, which enhances the photoelectrocatalytic breakdown of organic molecules. Kerkez *et al.* [64] worked to generate TiO_2 nanorod array film, which was placed on fluorine doped tin oxide (FTO) and was additionally CuO doped upto 0.26% in terms of Cu^{2+} molar ratio. The E_{bg} of doped photocatalysts decreases with increased proportion of the dopant, from 3.1 eV to 2.6 eV in TiO_2/FTO. This facilitates the photoelectrocatalytic system to function effectively under irradiation with the visible spectrum, enhancing degradation of organic contaminants. Another effective method of improving effectiveness of TiO_2 is through synthesis of composites modified in the presence of noble metals and additional semiconductors. Photoelectrocatalytic process can further be improved by doping TiO_2 nanotubes with Ag, Pd, and Au [65], which is mainly associated with the low work function of noble metals that initiates a Schottky blockade potential at the interphase. It facilitates transfer of electrons from CB of semiconductors toward metals, with reduction in recombination of light-induced E_{CB}^-/h_{vb}^+ pairs appreciably.

8.4 Applications of Modified Photocatalysts

8.4.1 Removing Organic Dyes Present in Wastewater

Industries use organic dyes and their release in water seriously damages the aquatic environment. An organic dye absorbs light and adheres firmly to a fiber through chemical or physical linkages. Typically, dyes broadly categorized on their chemical structure and the presence of

Table 8.1 Modified (doped and codoped) photocatalysts used for treatment of dyes (reproduced with permission) [70].

Photocatalyst	Dye	Concentration of dye (ppm)	Treatment time (min)	Percentage efficiency of treatment
Fe-TiO_2	Methylene blue	30	60	93
Mn-TiO_2	Methylene blue	10	180	97
Mn-TiO_2	Methylene blue	160	300	75
Sn-TiO_2	Methylene blue	20	120	77
Ni-TiO_2	Methylene blue	5	180	98
Ni-TiO_2	Methylene blue	10	60	93
Si-TiO_2	Methylene blue	10	120	87
Al-TiO_2	Rhodamine B	10	180	89
Al-TiO_2	Acid orange 7	5	60	99
Pt-TiO_2	Rhodamine B	10	90	99.5
Fe-TiO_2	Acid orange 7	10	60	73
Fe-TiO_2	Methylene blue	2	240	78
Pd-TiO_2	Methylene blue	20	120	99.4
C-TiO_2	Methylene blue	10	100	100
N-TiO_2	Acid orange 7	10	60	55
S-TiO_2	Rhodamine B	10	70	95
I-TiO_2	Rhodamine B	20	240	95
C-N-TiO_2	Rhodamine B	20	90	94
Fe-N-TiO_2	Acid orange 7	10	60	83
Fe-Pr-TiO_2	Acid orange 7	10	60	80
Fe-La-TiO_2	Methylene blue	0.1	180	44
Fe-Ce-TiO_2	Methylene blue	4	120	86
F-N-TiO_2	Methylene blue	5	300	97

distinct chromophores [66]. The chemical structure of dyes cannot be degraded by microbes. Dyes are extremely toxic organic molecules and very difficult to degrade [67]. They may cause irritation in the eye and skin, precipitate asthmatic attacks, contagious dermatitis, and infections of the respiratory tract [68]. Azo dyes can form aromatic amines under hazardous anaerobic conditions. Dyes block photosynthesis in aquatic plankton by slowing diffusion of light in water; hence, oxygen is not available for degradative processes of microbes in water [69]. The photocatalytic breakdown of dyes is reliant on *in situ* generation of hydroxyl radicals, and they are rapid in action and can oxidize any dye contaminant. Table 8.1 depicts the action of doped TiO_2 photocatalysts on dyes contaminating wastewater.

8.4.2 Removal of Emerging Contaminants (ECs) from Wastewater

Uncontrolled pollutants requiring jurisdictive procedures are called emerging contaminants (ECs). They are not newly introduced contaminants but are residing in the environment since a long period of time [71].

ECs are major constituents of various products used in everyday living, particularly the pharmaceuticals, endocrine-disrupting chemicals (EDCs), personal care products (PPCPs), and persistent organic pollutants (POPs) [72].

The concentration of ECs found in municipal wastewater, surface water, and water for drinking purposes is very low, but they have stable chemical character and are prone to accumulation. Hence, they are detrimental to health and environment [73, 74].

EDCs target specific organs, and a few psychotropic compounds lead to neurological damage [75].

Generally, ECs are toxic residues formed upon incorrect and repeated use of products in daily life [76]. These contaminants are disposed of or excreted by the inhabitants and ultimately go in wastewater treatment plants [77]. There are very few biodegradable residues derived from drugs released into surface water along with water from treatment plants. Besides, the drug remnants percolate into ground water through diffusion, runoff, or by infiltrating surface water, causing contamination of potable water. ECs are also found in the sludge from water treatment plants, which can be used in agriculture and cultural fertilization, further leading to contamination of surface waters. Recently, various studies have reported photocatalytic strategies for removing ECs [78]. Prieto-Rodriguez *et al.* revealed that TiO_2 in lower concentrations is effective for removing ECs from the exudants of wastewater treatment plants. Table 8.2 depicts the effect of modified TiO_2 as a photocatalyst, prepared for degradation of ECs [79].

Table 8.2 Effect of modified photocatalysts on degradation of ECs (reproduced with permission) [70].

Photocatalyst	**Emerging contaminant (EC)**	**Concentration of EC (ppm)**	**Treatment time (min)**	**Percentage efficiency of treatment**
Fe-TiO_2	Paraquat dichloride	30	75	98
Fe-TiO_2	Gesaprim herbicide	40	200	90
N-TiO_2	Mecoprop	580	1,200	50
Ni-TiO_2	Ibuprofen	25	360	50
Bi-TiO_2	Ibuprofen	25	360	89
Ce-TiO_2	Tetracycline	20	60	93
Mn-TiO_2	Glyphosate	6	300	75
N-TiO_2	Lindane	100	330	100
Fe-TiO_2	Lindane	100	330	10
Ag-TiO_2	Lindane	100	330	23
Cr-TiO_2	Lindane	100	330	30
Fe-Cr-TiO_2	Lindane	100	330	37
Fe-N-TiO_2	Streptomycin	30	240	50
Fe-N-TiO_2	Amoxicillin	30	240	59
N-S-TiO_2	Aldicarb pesticide	19	90	>80
Cu-S-TiO_2	Ciprofloxacin	15	90	100
C-N-B-F-TiO_2	Imidacloprid	24	450	71
C-N-B-F-TiO_2	Thiamethoxam	87.5	480	65
Fe-N-P-TiO_2	Thiachloprid	0.5	180	29

8.4.3 Photocatalytic Degradation of Phenols

Phenols have hydrogen substituted by hydroxyl groups on phenyl rings, which can be halogenated, alkylated, or nitrated to form a variety of related structures. The hydroxyl group in phenols links with water molecules through hydrogen bonding and is classified as endogenous and exogenous [80]. Endogenous phenols occur naturally and are usually present in leaves, roots, stems, fruits, and many other plant segments. Exogenous phenols are generated through combustion of biofuels and automobile exhaust. Phenols substituted with methoxy, methyl, carbonyl, and carboxylate are widespread. A significant study reported emissions from agricultural biomass combustion in the form of catechol and guaiacol, which are hazardous phenolic compounds [81]. Furthermore, manufacturing and a few industrial processes generates many phenolic compounds; for instance, they are major extracts from the refining processes carried out in oil and coal mines [82]. Domestic and industrial activities are known to release noxious phenolic compounds into water bodies [83], for example, chlorophenols are generated in paper manufacturing [84]; methylphenols, 2,4-dinitrophenol, and chlorophenols are commonly used in production of pesticides, preservatives, herbicides, and dyes [85]; whereas cyanophenols are employed to produce drugs and fungicides [86].

Phenols have weak acidic character, with a comparatively higher water solubility, requiring more time for degradation and have a higher tendency of bioaccumulation. Nitrophenols absorb UV and visible light, resulting in changed atmospheric photochemistry, affecting air quality and altering the climate. In soil or water, phenolic contaminants alter the pH, resulting in adverse effects on animals, plants, and micro-organisms inhabiting soil or water. Phenols are very toxic and can attack the body by way of the skin, mucosa, digestive tract, and respiratory tract, denaturing and coagulating proteins, resulting in loss of cellular activity. Chlorophenols and alkylphenols are specific endocrine disruptors, found in water and soil, and are detrimental to normal functioning of different organs, as they are embryotoxic and teratogenic [87]. Treatment with AOPs has been effectively employed for industrial wastewater reclamation, by generating strong reactive species, mainly the sulfate and hydroxyl radicals. AOPs can reduce the toxic character of phenolic compounds by breaking their structure and dismantling long chain molecules. Photo-Fenton (PF) and H_2O_2/UV-C treatment can effectively degrade 2,4-dichlorophenol, accompanied by removal of total organic carbon (TOC) [88]. Codoped TiO_2 effectively detoxicated phenols when irradiated with visible light [89, 90]. Table 8.3 depicts the action of doped and codoped TiO_2 photocatalysts on phenolic compounds upon irradiation with visible light.

Modifying TiO_2 semiconductor with metals and non-metals is essential for enhancing the photocatalytic abilities in many ways. For example, the prospect of employing light sources besides UV (for minimizing cost and enhancing safety) increases with doped and codoped semiconductors. Noble metals enhance photocatalytic efficiency of TiO_2, on irradiation of both ultraviolet and visible spectra. Noble metals induce changes on the exterior of semiconductor, but the crystal lattice remains unchanged [91].

Table 8.3 Doped and codoped TiO_2 photocatalysts targeting phenolic compounds [70].

Photocatalyst	Phenolic compound targeted	Concentration of phenolic compound (ppm)	Treatment/ exposure time (min)	Percentage efficiency of treatment
Ag-TiO_2	4-Chlorophenol	10	360	73.5
N-TiO_2	Phenol	50	120	100
N-TiO_2	Phenol	20	120	36
N-TiO_2	Phenol	25	600	81
N-TiO_2	4-Chlorophenol	130	600	17
Fe-TiO_2	Phenol	10	210	30
Fe-TiO_2	Phenol	5	240	40
Ce-TiO_2	2-Chlorophenol	10	240	100
Sm-TiO_2	4-Chlorophenol	130	600	1.5
Sm-N-TiO_2	4-Chlorophenol	130	600	40
Cr-V-TiO_2	Phenol	5	180	50
N-Zr-TiO_2	4-Nitrophenol	14	1400	68
Fe-Pr-TiO_2	Phenol	10	240	66
Pt(II)-Cr-TiO_2	Phenol	5	180	70

The above changes do not alter the bandgap; hence, absorption potential of light sources except UV remains unchanged, due to blockage in the recombination of electron–hole pair. This maintains stability of semiconductors like TiO_2 and their effectiveness as a photocatalyst [92]. When non-metals are utilized, they usually enter the crystal lattice and replace an atom comparable in size, reacting with oxygen atoms and altering the electron density on them [93–95]. This changes the energy levels of the oxygen atom, consequently modifying the bandgap of the photocatalyst and facilitating absorption of light at different wavelengths [96]. Codoping strategies improve the electronic, chemical, and magnetic characteristics of the original semiconductor [97]. Research confirms that, in comparison to a single-doped TiO_2, codoping enhanced the efficiency of TiO_2 to absorb visible light [98]. Codopants not only reduce the bandgap but also contribute toward formation of various heterostructures, for example, different electronic structures with TiO_2 [99].

8.4.4 Anti-Microbial Efficacy of Treatment by Photocatalysts

Efficient antimicrobial treatment of wastewater is dependent on concentration and type of photocatalyst, the pathogen present, pH, temperature, level of hydroxylation, magnitude of radiation, and the time of exposure [100, 101]. Effectiveness of antimicrobial treatment covers a variety of outcomes, which can be assessed by analyzing one study at a time. The most frequently employed procedure is through measurement of the percentage reduction in original microbial count after treatment.

Many modified photocatalysts are effectively designed for microbicidal action, for example, development of heterostructures (Table 8.4), doped structures (Table 8.5), and nanocomposite polymers (Table 8.6). A new area of activity is expansion of TiO_2 nanotubes and their pairing with metal oxides, cations, and other compounds, enhancing excitation in the visible spectrum [102, 103].

Table 8.4 Antimicrobial action of heterostructured photocatalysts [104].

Photocatalyst	Microbe targeted	CFU/mL	Treatment time (min)	Source of light	Treatment efficiency (%)
ZnO/TiO_2	*E. coli*	10^7	10	Visible	99.9
$Au/BiTiO_3/TiO_2$	*E. coli*	6×10^4	40	Simulated solar radiation	99.5
	S. aureus				99.7
$g\text{-}C_3N_4\text{-}V\text{-}TiO_2$	*E. coli*	-	60	Visible	99.5
	S. aureus				
$Ag_2C_2O_4/TiO_2$	*E. coli*	2×10^7	30	Visible	99.99
$Ag/BiOI/TiO_2$	*E. coli*	3×10^7	30	Visible	99.9
$CuBi_2O_4/Bi_2MoO_6$	*E. coli*	10^7	240	Visible	100
$Ag_2WO_4/g\text{-}C_3N_4$	*E. coli*	10^7	90	Visible	100
Ag QDs/$Bi_2S_3/SnIn_4S_8$	*E. coli*	2.5×10^7	240	Visible	100
$Fe_2O_3\text{-}TiO_2$	*V. fischeri*	3×10^6	30	UV	100
MoS_2/TiO_2	*S. aureus*	$>10^8$	150	Visible	100
	E. coli				

(Continued)

Table 8.4 Antimicrobial action of heterostructured photocatalysts [104]. (*Continued*)

Photocatalyst	Microbe targeted	CFU/mL	Treatment time (min)	Source of light	Treatment efficiency (%)
TiO_2-rGO	*E. coli*	1.5 x 10^6	75	Artificial solar light	100
g-C_3N_4-AgBr	*S. aureus*	3 x 10^6	150	Visible	100
	E. coli		60		
AgI/AgBr/ $BiOBr_{0.75}I_{0.25n}$	*E. coli*	3 x 10^7	30	Visible	100
Al_2O_3/ZnO	*E. coli*	10^6	240	Visible	100
g-C_3N4/expanded perlite (EP-520)	*E. coli*	1 x 10^8	120	Visible	100
	MS2		240		
TGP (TiO_2–graphene sensitized by tetrakis(4-carboxyphenyl) porphyrin (TCPP))	*E. coli*	-	440	Visible	64
Ag-AgX/RGOs	*E. coli*	2 x 10^7	35	Visible	100
AgI@MnO_2	*S. aureus*	-	25	Visible	99.4
	E. coli				92.2
CeO_2-AgI	*E. coli*	10^7	40	Visible	100
Ag-AgX/RGOs	*E. coli*	2 x 10^7	35	Visible	100
O-g-C_3N_4/HTCC-2	*Viruses*	10^5 MPN/ ml	120	Visible	100
$BiVO_4$/Ag^+	*E. coli*	10^8	15	Visible	>99
BiOBr-AgBr	*E. coli*	1 x 10^7	24	Visible	100
TiO_2–Fe_2O_3	*E. coli*	3.22 x 10^9	120	Natural sunlight	98.3
$Ag_{(3\%)}$-TiO_2	*E. coli*	10^6	60	Natural sunlight	100
GO/g-C_3N_4	*E. coli*	10^7	120	Visible	100
TiO_2-X/Ag_3PO_4	*S. aureus*	10^7	20	Simulated sunlight	99.8
	E. coli				99.8

Table 8.5 Antimicrobial action of doped photocatalysts [104].

Photocatalyst	Microbe targeted	CFU/mL	Treatment time (min)	Source of light	Treatment efficiency (%)
N-TiO_2	*E. coli*	10^5	360	Visible	-
Cr-TiO_2					70
Cr/N-TiO_2					-
TiON	*E. coli*	10^6	40	Simulated light	100
TiO_2-Cu	*E. coli*	3.8×10^6	120	Visible	100
N-TiO_2	*E. coli*	2×10^6	240	UV	33
N-TiO_2	*Aspergillus niger*	10^5	7,800	Visible	100
N-T-TiO_2			7,200		
C-TiO_2			7,200		
Pd-C-TiO_2			5,760		
TNTZ-Cu	*E. coli*	3×10^6	75	Visible	100
TiO_2/CdS	*E. coli*	10^8	10	Visible	99
TiO_2/Cu	*E. coli*	10^6	10	Simulated solar radiation	100
V_2O_5/TiO_2	*E. coli*	10^8	30	UV-C Visible	100
TiO_2/Cu-PES	*E. coli*	$>10^6$	30	Actinic light	100
PECuOx	*E. coli*	$>10^6$	15	Sunlight	100
Ce-ZnO	*E. coli*	1×10^5	120	UV-A	99.99
	P. aeruginosa				
Fe-TiO_2	*E. coli*	10^6	120	Sunlight	>99
F-ZnO	*S. aureus*	-	360	Visible	99.99
	E. coli				99.87
TiN/TiN-Ag	*E. coli*	10^8	15	Actinic light	100
CuOx-TiO_2-PET	*E. coli*	4×10^6	20	Actinic light	100
Ti- BiOI	*E. coli*	3×10^7	24	Visible	100
	S. aureus	3×10^6	45		

(*Continued*)

Table 8.5 Antimicrobial action of doped photocatalysts [104]. (*Continued*)

Photocatalyst	Microbe targeted	CFU/mL	Treatment time (min)	Source of light	Treatment efficiency (%)
$ZnCl_2/TiO_2$,	*S. aureus*	10^5–10^6	120	Visible	>90
$Zn(Ac)_2/TiO_2$,					>80
$Zn(NO_3)_2/TiO_2$					>95
$ZnSO_4/TiO_2$					100
$ZnCl_2/TiO_2$,	*E. coli*	10^5–10^6	120	Visible	>92.5
$Zn(Ac)_2/TiO_2$,					>80
$Zn(NO_3)_2/TiO_2$					>90
$ZnSO_4/TiO_2$					100
$ZnCl_2/TiO_2$,	*Candida albicans*	10^5–10^6	120	Visible	>95
$Zn(Ac)_2/TiO_2$,					>87.5
$Zn(NO_3)_2/TiO_2$					>87.5
$ZnSO_4/TiO_2$					100
Ce-ZnO	*E. coli*	10^6	120	UV-A	100
Cu-ZnO	*E. coli*	$10^{6.5}$	240	Simulated solar radiation	100
ZnO/TiO_2	*E. coli*	10^5	20	UV	100

Coupling Zn with TiO_2 enhances efficiency of the photocatalyst upon irradiation of solar and UV light, as observed in studies on heterostructures and doped compounds (Tables 8.4 and 8.5). According to a recent study (Table 8.5), doped TiO_2/ZnO effectively removed 105 CFU/mL *E. coli* after 20 min of treatment, in the presence of UV light. Another study reported similar results (Table 8.5) after irradiation with visible light and only 10 min of treatment [101]. $ZnCl_2/TiO_2$ and $Zn(Ac)_2/TiO_2$ nanoparticles used visible light and removed *C. albicans* more efficiently than *E. coli* and *S. aureus* [105] (Table 8.5).

TiO_2-Fe_2O_3 nanocomposites effectively remove *E. coli* (99% extrication) on exposure to direct sunlight, although demanding higher treatment time of approximately 120 min [106] in comparison to exposure under UV light, where effective treatment time was 30 min [107].

Table 8.6 Antimicrobial action of polymer-based nanocomposite photocatalysts [104].

Photocatalyst	Microbe targeted	CFU/mL	Treatment time (min)	Source of light	Treatment efficiency (%)
Chitosan-TiO_2:Cu (CS-CT)	*E. coli*	3×10^4	120	Visible	100
	S. aureus		150		
PMMA/TiO_2	*E. coli*	10^5	60	UV-A	70
PMMA/TiO_2/SWCNTs				UV-A	-
PMMA/TiO_2-TCPP				Visible	40
Ag-NPs@CTA	*E. coli*	10^8	120	Visible	99
	S. aureus				
	C. albicans				

Additionally, many more photocatalysts that are visible light driven (VLD) have been investigated, for instance, Bi-containing [108–110], Ag-containing [111–113], and C-containing photocatalysts. Bismuth oxyhalide (BiOI) semiconductor exhibited maximum visible light absorption, owing to a narrow bandgap with 1.7 eV to 1.8 eV. Heterostructures shaped by coupling TiO_2- and BiOI-coupled heterostructures facilitate the effective removal of microbes under visible light. Although, an incorrect match in band alignment within TiO_2 and BiOI restricts the interfacial charge transmission. To remove this anomaly, a recently conducted study combined TiO_2/BiOI nanoparticles with Ag nanoparticles, which effectively enhanced absorption of light and improved charge transfer [112]. As depicted in Table 8.4, Ag-TiO_2/BiOI photocatalyst, effectively removed 3×10^7 CFU/mL of *E. coli*, after 30 min of treatment time on exposure to visible light.

Another semiconductor, molybdenum disulfide (MoS_2), is also utilized to enhance the range of TiO_2 on visible light exposure and refine effectiveness of charge separation [103]. Graphitic carbon nitride (g-C_3N_4) is another innovative photocatalyst used for the removal of microbes [114, 115]. g-C_3N_4 codoped with V-TiO_2 [116], AgBr [117], and expanded perlite [115] exhibit effective antimicrobial properties against *E. coli* under visible light (Table 8.4), but no effects are reported with solar radiation. In recent years, many 2D designed semiconductors and their composites [119], which include nanocomposites based on Ag and graphene oxide

(GO), have been developed as efficient antimicrobial photocatalysts [120, 121].

g-C_3N_4 coupled with graphene oxide (GO/g-C3N4) generated a heterostructure, which removes 97.9% of *E. coli* at concentration 100 μg/mL, post 120 min of irradiation with visible light (Table 8.4). A variety of ternary photocatalysts modified with Ag, like Ag/QDs/BiS_3/$SnIn_4S_8$ [122], Ag-AgX/RGOs [123], and AgI/AgBr/BiOBr [108], exhibit enhanced photocatalytic activity on exposure to visible light, which is mostly associated with surface plasmon resonance (SPR) and Schottky effect exhibited by Ag nanoparticles. Most effective photocatalytic process under visible light was reported in a study with lowermost concentration of photocatalytic semiconductors and the maximum concentration of *E. coli* at 3×10^7 CFU/mL [108].

8.4.5 Innovative Catalysts for Removal of Herbicides

Herbicides can be degraded biologically by microorganisms, bacteria, or fungi, in the presence or the absence of oxygen. This microbial treatment lowers cost and is a favorable method to remove contaminants [124]. However, the microorganisms required act selectively on specific herbicides, are temperature sensitive, and, sometimes, lack longevity [125]. Process of adsorption uses adsorbents that have a high surface area for trapping contaminants on the surface or within pores. Adsorption has a major disadvantage as the chemical pollutants are not mineralized and the used adsorbants need decontamination before disposal [126]. Membrane filtration, another treatment process, removes herbicides, with a size more than the size of membrane pores [127]. In saline wastewater, the traditional processes of reverse osmosis (RO) and nano filtration (NF) are employed for removing 2-methyl-4- chlorophenoxy acetic acid (MCPA), exhibiting effectiveness of 95.3% and 71.3%, respectively [128]. However, the membrane processes face a common problem of membrane fouling, which delimits the efficiency of membrane function. Regular cleaning is required which inflates operating cost and releases additional waste material [129].

Photocatalysis is conducive for herbicide removal as there is no further requirement of additional chemicals to degrade organic contaminants [130]. Studies on use of photocatalysts in herbicide treatment indicate a steady increase since the year 2000. Although the procedures for developing a photocatalyst are common, the factors required for degrading herbicides and other contaminants vary [20]. This section elaborates the recently developed photocatalytic design for removal of herbicides in aquatic environment.

Another study reported improved degradation of 4-chlorophenoxyacetic acid (4-CPA) by doped N-TiO_2 photocatalyst on exposure to visible light. The increase in decomposition of herbicide by N-TiO_2 was caused by reduction in size and bandgap energy of the photocatalyst [131]. Other investigators also confirmed that a decrease in size ensures an increase in precise area of the crystalline photocatalyst, which enhances effectiveness of the photocatalyst as more sites are available for action [132]. The consequence of temperature and aging on synthesis of N-TiO_2 was studied for removal of herbicide alachlor [133].

Another strategy supported N-TiO_2 with magnetic hexagonal mesoporous silica (magnetic HMS) for degradation of 2,4-D, trifluralin, and glyphosate herbicides on exposure to visible and UV light [134]. Magnetic HMS reduced agglomeration of the photocatalyst and played the role of an adsorbent for herbicide removal. Photocatalytic complex is conveniently recycled after the reaction without appreciable harm, owing to the ferromagnetic properties of HMS.

In another study, degradation of atrazine was assessed, using TiO_2 doped with boron, which was shown to be four times more effective than the unmodified TiO_2. Enhanced photocatalytic activity depended on the tuning of rutile TiO_2 material, so that it functions as a trap for improved separation of charge carrier.

Furthermore, doping with boron increased the photocatalytic TiO_2 oxidation by an extension of VB toward positive character, which facilitates hydroxyl radical generation [135]. The effectiveness of N-ZnO films in mineralization and breakdown of 2,4-D was studied on exposure to solar radiation [136]. N-ZnO thin films were generated through the radiofrequency magnetron sputteriprocedure, by which the highly purified ZnO was deposited over a glass substrate by evaporating it simultaneously with Ar gas–based plasma and gaseous N_2. N doping acted as a protective shield against corrosive effects of light, with the thin films exhibiting consistent effectiveness over a period of time. A few research groups reported that non-metallic doping would result in generation of another level near conduction band or valence band, causing increase in photo-absorption under visible light and decreases recombination of charge carriers [22, 28, 133, 137], but a study reported diversely. After successfully doping C and N on porous nanospheres of Bi_2O_3, the researcher found that there was no change in the valence band of modified Bi_2O_3. Although, the doped photocatalysts degraded atrazine more effectively in comparison to the undoped Bi_2O_3 [138]. These results can be explained by various ways. Firstly, the dopants may act as surface sensitizers, which, upon irradiation with visible light, transfer electrons to CB. Another explanation can be related to C and

N, which create an energy level constituting their 2p orbitals, in vicinity of the CB of Bi_2O_3, which causes a double step electron shift from VB to CB of Bi_2O_3 through the new energy levels. Thirdly, the recombination of charge carriers can be restrained by dopants through electron trapping, thereby holding back the process of recombination.

Along with non-metals, metals are also being used for doping to enhance the photocatalytic degradation of herbicides. Use of Fe^{3+} in doping TiO_2 has been observed in degradation of atrazine on exposure to UV irradiation. The doped TiO_2 degraded atrazine with an effectiveness of 99.47%, at pH 11 in a time frame of 30 min, when the catalyst is loaded at concentration 25 mg L^{-1}, owing to a high concentration of hydroxyl ions in alkaline conditions [139]. In another investigation, the role of Cr(III)-doped TiO_2 in photocatalytic degradation of MCPA (2-methyl-4-chlorophenoxyacetic acid) herbicide was observed on exposure to UV and visible spectra. The study reported that Cr(III)-doped TiO_2 was capable of fully degrading MCPA on exposure to visible and UV radiation within 240 min and 150 min, respectively. These researchers further confirmed that Cr(III) doping may block the growth of TiO_2 crystals and initiate deformities in the lattice, efficiently increasing separation of electron–holes [140]. Another research used Cu-doped ZnO nanorods for bentazon degradation, a 100% degradation was observed in 60 min of treatment, at pH 7.0, and on UV exposure [141]. ZnO doped with Cu was highly effective as Cu^{2+} ions trapped the photo-generated electrons, thereby increasing the separation of electron-hole pairs. Solution pH plays a critical role in enhancing efficiency of a photocatalyst. At pH more than 3, bentazon was negatively charged, whereas the Cu-doped photocatalyst was positively charged, resulting in an electrostatic attraction. Therefore, bentazon was attracted with force to the surface of doped photocatalyst and was able to complete the photocatalytic reaction. Table 8.7 lists several photocatalysts employed for herbicide degradation.

Accumulation of noble metals on surface of semiconductors enhances the efficiency of a photocatalyst by ameliorating the response to visible light and through decreasing the recombination of electron–hole pairs [142, 143]. Another group synthesized Ag-ZnO and discovered that the decorated ZnO degraded diuron herbicide at a rate which was 14 times higher than pristine ZnO on exposure to solar radiation. The bandgap energy of ZnO decreased from 3.22 eV to 3.02 eV on integrating 0.7% Ag with ZnO [144].

For atrazine when treated with Ag-loaded $LaTiO_3$ nanowires, as reported by Shawky *et al.*, a Ag loading of 2.5 wt% showed the maximum photocatalytic efficiency, when it completely degraded the herbicide in

Table 8.7 A list of metal/non-metal-based photocatalysts for herbicide degradation [40].

Photocatalyst	Herbicide degraded	Herbicide concentration	Treatment time (min)	Source of light	Treatment efficiency (%)
S-TiO_2	2,4-Dichloro-phenol (2,4-DCP)	25 mg L^{-1}	240	Visible	91
N-TiO_2	Alachlor	20 ppm	180	UV	73.97
N-TiO_2/ magnetite HMS	Glyphosate	-	80	UV	87.9
				Visible	17
B-TiO_2	Atrazine	10 mg L^{-1}	180	Visible	94.7
Fe^{+3}-TiO_2	Atrazine	10 mg L^{-1}	30	UV	99.47
Cr (III)-TiO_2	4-Chloro-2-methylphenoxyacetic acid (MCPA)	10 mg L^{-1}	240	UV	100
				Visible	100
WO_3-TiO_2	2,4-Dichlorophenoxyacetic acid (2,4-D)	50 mg L^{-1}	120	UV	50.5
				Visible	59.2
				Solar	88.6
Fe_2O_3-TiO_2	2,4-Dichlorophenoxyacetic acid (2,4-D)	50 mg L^{-1}	120	UV	14.4
				Visible	43.6
				Solar	96.8
Ca_xMnO_y/ TiO_2	Imazapyr	5 μM	60	Visible	100
N-ZnO thin films	2,4-Dichlorophenoxyacetic acid (2,4-D)	20 ppm	240	Solar	94
C/N doped Bi_2O_3	Atrazine	5 mg L^{-1}	80	Visible	>70
W-ZnO	Glyphosate	20 mg L^{-1}	180	Visible	74

40 min of treatment time on exposure to visible light. This excellent photocatalytic activity was due to factors like enhanced visible light absorption and lower bandgap energy, with inhibition of light generated charge carriers. Furthermore, efficiency continued even after several photodegradation cycles.

Table 8.8 Photocatalysts doped with noble metals for degradation of herbicides [40].

Photocatalyst	Herbicide degraded	Herbicide concentration	Treatment time (min)	Source of light	Treatment efficiency (%)
Ag/ZnO	Diuron	10 mg L^{-1}	90	Solar	100
Ag $LaTiO_3$	Atrazine	50 ppm	40	Visible	100
Pd-WO_3 nanorods	2,4-Dichlorophe-noxyacetic acid (2,4-D)	40 ppm	50	Visible	100
Pt/TiO_2–SiO_2	2,4-Dichlorophe-noxyacetic acid (2,4-D)	50 mg L^{-1}	180	UV	97
Pt-$BiFeO_3$	2,4-Dichloro-phenol (2,4-DCP)	5 mg L^{-1}	240	Visible	100

Khalid synthesized WO_3 nanorods doped with Pd for removal of 2,4-D on exposure to visible light. The author reported that 0.15 wt% was most efficient, exhibiting 100% degradation efficiency, in a treatment time of 50 min [145].

In another study, Pt was doped on TiO_2-SiO_2 to increase decomposition of herbicide 2,4-D [146].

Photocatalytic breakdown of 2,4-DCP with $BiFeO_3$ doped with noble metals, on exposure to visible light, was efficient as reported by Lam *et al.* [147].

The significant results of selected investigations confirming removal of herbicides by employing noble metal doped photocatalysts are given in Table 8.8.

8.5 Solar/Photoelectro-Fenton Process

In Fenton's reaction, hydrogen peroxide is converted into a hydroxyl free radical through catalysis [148]. The hydroxyl free radical formed is unstable and extremely toxic. The reagent combines Fe^{2+} and H_2O_2, which generates homogeneous hydroxyl radical ·OH [149]. The major limitation of this reaction is formation of Fe^{+3} and short-chain aliphatic carboxylate complexes, which cannot be oxidized by ·OH radical. This limitation can be overcome by employing the PF or solar PF (SPF) methods, in which the contaminant is irradiated with ultraviolet or solar light to generate ·OH

Table 8.9 Selected studies on effectiveness of homogeneous photoelectro-Fenton (PEF), solar PEF (SPEF), and PEF-like processes on degradation of several organic contaminants in water bodies [148].

	Organic contaminant targeted	**Concentration of contaminant**	**Treatment time (min)**	**Source of light**	**Treatment efficiency (%)**
Homogeneous PEF/PEF-like processes					
Industrial waste	2,6-Dimethyl aniline	1 mM	120	UV-A	100
	Dimethyl sulfoxide	10 mM	120	UV-A	100
	Phenol	2 g L^{-1}	360	UV-C	85% TOC removal
	Methylparaben	1.04 mM	360	UV-A	100
	Trans-Ferulic acid	0.834 mM	360	UV-A	100
Herbicides	Alachlor	0.60 mM	360	UV-A	100
	Cyanazine	110 mg L^{-1}	240	UV-A	100
	Imidacloprid	20 mg L^{-1}	300	UV-C	100
	Carbofuran	0.348 mM	240	UV-A	100
	Metolachlor	0.55 mM	360	UV-A	100
	Metribuzin	0.523 mM	420	UV-A	100
Dyes	Acid Red 29	244 mg L^{-1}	180	UV-A	100
	Direct Red 23	30 mg L^{-1}	60	UV-C	90
	Acid Blue 5	30 mg L^{-1}	60	UV-C	100
	Orange G	295 mg L^{-1}	120	UV-A	100
	Reactive Black 5	158 mg L^{-1}	240	UV-C	100
Drugs	Sulfamethoxazole	200 mg L^{-1}	360	UV-A	100
	Flumequine	62 mg L^{-1}	360	UV-A	100
	Acetaminophen	5 mM	120	UV-A	99
	Dopamine	1.04 mM	240	UV-A	97
	Fluoxetine	0.490 mM	360	UV-A	100
	Ciprofloxacin	200 mg L^{-1}	360	-	100
	Cefalexin	200 mg L^{-1}	360	-	100

(Continued)

Table 8.9 Selected studies on effectiveness of homogeneous photoelectro-Fenton (PEF), solar PEF (SPEF), and PEF-like processes on degradation of several organic contaminants in water bodies [148]. (*Continued*)

	Organic contaminant targeted	Concentration of contaminant	Treatment time (min)	Source of light	Treatment efficiency (%)
Real waste-waters	Coffee wastewater	4,200 mg L^{-1}	240	UV-A	92
	Slaughterhouse effluent	1159 mg L^{-1}	60	UV-A	92
	Tetracaine in urban wastewater	0.561 mM	360	UV-A	100
Homogeneous SPEF					
Industrial chemicals	Phthalic acid	2.0 mM	240	Direct sunlight	100
	Methyl, ethyl, and propyl paraben in urban wastewater	0.30 mM	240	Solar energy	100
Herbicides	4-Chloro-2-methylpheno-xyacetic acid	186 mg L^{-1}	120	Solar energy	100
	Diuron	0.185 mM	240	Solar	100
	Tebuthiuron and Ametryn	0.18 mM	360	Solar	100
Dyes	Sunset Yellow FCF	290 mg L^{-1}	270	Solar	100
	Direct Yellow 4	290 mg L^{-1}	240	Solar	100
	Allura Red AC	460 mg L^{-1}	360	Solar	100
	Acid Yellow 42	284 mg L^{-1}	270	Solar	100
	Acid Blue 29	118 mg L^{-1}	300	Direct sunlight	100
Drugs	Sulfanilamide	239 mg L^{-1}	180	Solar	100
	Chloramphenicol	245 mg L^{-1}	180	Solar	100
	Trimethoprim	200 mg L^{-1}	360	Solar	100
	Levofloxacin	60 mg L^{-1}	360	Solar	100
Real waste-waters	Washing machine effluent	315 mg L^{-1}	240	Direct sunlight	100

radicals and photolyze the Fe^{+3}-carboxylate complexes, thereby significantly increasing degradation [150, 151]. Electrochemical AOPs (EAOPs) trigger great enthusiasm as they are environmentally compatible, requiring moderate conditions and are safe, are effective, and have the convenience of automation [152, 153]. The least complicated EAOP dependent on Fenton's reaction is called electro-Fenton (EF), exhibiting improved degradation ability. EF involves an onsite production of Fenton's reagent (Fe^{2+} and H_2O_2) at the cathode, where the removal of organic compounds can happen not only in the presence of homogeneous ·OH radical generated in Fenton's process but also by direct oxidation at anode or by oxidation mediated by heterogeneous ·OH radical generated from anodic water splitting [154]. It is an improvement over the simple conventional anodic oxidation (AO), where degradation occurs by the latter processes of AO, which can be direct or mediated [155].

Initial studies on application of EF in wastewater reclamation were done by the Brillas and Oturan groups from 1994 to 2001, reporting rapid degradation of organic moieties, although, in a few cases, low levels of mineralization were observed due to ineffective action of homogeneous and heterogeneous ·OH radical on non-degraded Fe^{+3} carboxylate complexes [149, 152, 153].

Irradiation of the EF solution with UV-A, as given in 1995 by Brillas *et al.* improved the effectiveness of mineralization through PEF process. In 2007, they employed solar radiation as a cost effective and renewable source in the SPEF process, which ameliorated the efficacy of PEF process through a stronger irradiance of the solar energy as compared to the conventional UV-A sources. In recent years, both PEF and SPEF processes have been studied by many researchers, and their effectiveness in the removal of organic waste from wastewaters has been successfully corroborated (Table 8.9).

8.6 Photocatalytic Fuel Cells (PFCs) and Their Relevance in Minimizing Use of Electrical Procedures

Photocatalysis is photo-induced combustion of organic compounds, and, when executed in an isolated chamber, the final process releases a similar electron flux as an additional advantage. These redox reactions are carried out in a cell, employing suitable electrolytic equipment. The significant factors are regulation of electric loss and choosing the right electrodes [156].

PFC is a process by which removal of organic pollutants and conservation of energy occurs simultaneously, by utilizing the photocatalytic reactions [157]. PFC has a photoanode, which is a photocatalyst, where the contaminants are oxidized under UV irradiation; and the electrons generated are carried toward the cathode, where redox reactions take place, completing the external circuit and generating electricity [158]. Owing to a large bandgap energy of pristine TiO_2 along with requirement of UV irradiation, several innovative photoanodes, capable of responding to visible light, have been developed in recent years, which have a wide absorption for the solar absorption, enhancing energy generation; for example, CdS/TiO_2 [159], CdS/ZnS/TiO_2 [160], $BiVO_4$/TiO_2 [161], $BiVO_4$/WO_3 [162], WO_3/W [163], and Ag/AgCl/TiO_2 [164]. To enhance the effectiveness of charge separation and energy generation, a combined double photoelectrode system is used, which has a photoanode and a photocathode, where an internal bias potential is generated. Examples of photocathodes employed are Cu_2O/Cu [165], ZnO/CuO [161], Pt-modified buried junction silicon [163], etc. CuO is another semiconductor possessing a low bandgap; therefore it is preferred for use as a photocathode [166, 167]. A few investigations studied the amalgamation of electro-Fenton reactions at the cathode with the reactions at photoanode, in order to increase effectiveness of oxidation by employing carbon-based electrodes like Fe-Fe_2O_3/carbon felt [157] and Pt-doped buried junction silicon [163]. A traditional single and double chamber (H-type) is used for performing the PFC process [168, 169]. Micro-PFC process, without membrane, has a lateral orientation of electrodes that permit simultaneous illumination of light and is feasible for practical application of the system [170, 171].

8.7 Photo Electro Peroxone Technique

Peroxone method of wastewater reclamation is an amalgamation of hydrogen peroxide (H_2O_2) and ozone (O_3) and hence abounds in free radical generation. Free radicals oxidize organic compounds in wastewater, but use of hydrogen peroxide increases the cost and creates storage and transport issues [172]. In addition, peroxone has limitations concerned with a low rate of oxygen-ozone conversion, which decreases in neutral and acidic conditions [173]. Photoelectro-peroxone (PEP), a hybrid electrolytic procedure, involves ozonation along with photolysis under UV irradiation, which is an encouraging tool used for removing recalcitrant organic toxicants [174, 175]. PEP employs oxygen and electricity as major sources of reactive oxygen species like H_2O_2 and free hydroxyl radical.

Peroxone reagents (H_2O_2 and O_3) are triggered on exposure to UV radiation, without requiring catalytic mediation. If any unused H_2O_2 and O_3 remains, it is conveniently hydrolyzed to H_2O and O_2. Furthermore, the process of PEP diminishes the unsought generation of toxic by-products, which may require additional procedures for treatment. Hence, PEP facilitates development of cleaner processes for wastewater treatment [176]. H_2O_2 used in PEP treatment is generated by action of two electrons at the surface of carbon cathode as given by Equation (8.7), a mixer continuously pumps O_2 and O_3 into the reactants, with the aid of an external ozone generator [177].

$$O_2 + 2H^+ + 2e^- \rightarrow H_2O_2 \tag{8.7}$$

At low pH, the reaction rate decreases. Fe^{+2} or F^{+3} ions can produce metal organic complexes with carboxylates, which cause a reduced mineralization of the contaminants [176]. To further enhance generation of free radical species and degradation of contaminants, electro-peroxone process was amalgamated with UV exposure, which induced photolytic decomposition of O_3 and H_2O_2, along with degradation of the metal organic complexes [178]. The above treatment does not generate a secondary contamination as the remnants are easily decomposed to H_2O and O_2.

8.8 Treatment of Real Wastewater Treatment by Photoelectrochemical Processes

8.8.1 Treatment of Landfill

A landfill is deposited by the physicochemical and biological degradation of waste and also by precipitating and percolating through the waste [179]. The leachates can enter and contaminate the groundwater; hence, landfill treatment is essential. The features of a landfill are dependent on the climate, age, and the type of waste [180]. Typically, landfill contains a heavy amount of organic contaminants, heavy metals, salts, nitrogen derived from ammonia, etc. [181]. Relatively younger landfills, which are <5 years old, are capable of liberating leachates of mainly biodegradable organic contaminants, possessing a biochemical oxygen demand (BOD)/chemical oxygen demand (COD) ratio greater than 0.3 and a COD greater than 10,000 mg L^{-1}. Whereas, older landfills aged >10 years degrade leachates composed of mostly recalcitrant humic and fulvic acids, exhibiting a BOD/

COD ratio of less than 0.1, at alkaline pH, with a lower COD of less than 4,000 mg L^{-1} [182].

Conventional oxidative techniques for wastewater treatment are not successful with leachates, particularly those from older landfills.

Older landfills with a highly refractory character, a BOD/COD of 0.004, a low COD of 5,500 mg L^{-1}, and pH 8 were included in a photoelectrochemical process [183]. The researchers made a comparative analysis of photolysis, electrolysis, heterogeneous photocatalysis, and photoelectrolysis, employing a photoelectrochemical reactor with a UV mercury lamp. The effectiveness of degradation was facilitated by the indirect electrochemical oxidative processes, in the presence of OH, H_2O_2, and chlorine, which were modified into radicals by the photochemical processes. They discovered that dark color of leachate restricted the process, necessitating requirement of a few coagulative strategies prior to EAOPs. The process of electrocoagulation (EC) can be considered an appropriate procedure of pretreatment to EAOPs, after which the color and COD removal is possible to some extent. The processes EC, EO, and UV-based $CuFe_2O_4$ AOP, when sequentially conducted, eliminated COD by 95.6%, BOD by 91.6%, TOC by 90.5%, and ammonia by 99.8% (NH_4-N). The effectiveness of the treatment procedure was established by the biodegradation test and phytotoxicity analysis [184]. Use of PEF also works for treatment of the landfill [181]. A statistical experimental design was devised to study the effectiveness of PEF process for treating landfill leachate [185]. Major features like pH, COD, ammonia nitrogen, and chloride were reported as 8.10, 2,747 mg L^{-1}, 3,348 mg L^{-1}, and 3,583 mg L^{-1}, respectively. Almost 90% color and 75% COD were removed after 90 min of treatment, under UV irradiation, with anode of Ti/RuO_2 -IrO_2 and a cathode of stainless steel, H_2O_2 at 153.2 mM, and Fe^{2+} at 39 mM concentrations. Although photo-EAOPs give excellent efficiency of degradation, employing a single technique for detoxifying acute contamination and cellular toxicity of by-products was more feasible [186]. Merging EF and PF processes in a PEF reactor, which constitutes anodes of boron-doped diamond (BDD) soft iron at current 0.9 A, 300 mg L^{-1} H_2O_2, and a system flow rate of 0.6 L/min, enhanced the process of decontamination. A 15-min treatment with PEF and a subsequent activated sludge process of 72 h, degraded 77.9% of COD, and decreased genotoxicity significantly.

Efficacy of PEF and biological oxidation combination has been studied by Seibert *et al.* (2019) [187]. Decrease in by-products was reported when PEF treatment of 45 min was followed by 24 h of biological oxidation. The efficacy of photo-electrolysis post–biological treatment was studied after a sequential anaerobic and aerobic treatment in a sequencing batch reactor, where efficacy of COD and ammonia nitrogen removal was analyzed in an electrolytic reactor (DSA-based) under UV irradiation [180]. The degradation was mediated by an indirect oxidative process, dependent on an *in situ* electro-induced active chlorine and radical moieties. Removal of almost 87% NH_3-N and 77% COD was accomplished after 8 h of treatment, at pH 5.0 and current 60 mA cm^{-2}, without adding more electrolyte. Another study analyzed the efficacy of photoelectrochemical treatment in a biologically pre-treated leachate, conducted with a capacity of 6.5 L in a continuous flow reactor at pilot scale. After 2.5 h of treatment, 41.6% TOC, 74.1% COD, and 94.5% ammonia were eliminated on application of 67.1 mA cm^{-2}. Ultraviolet irradiation led to mineralization of organic compounds to molecular acids, with a decrease in metal ions also reported through indirect oxidation.

8.8.2 Oil Mill Effluents

By-products of vegetable oil industry are acidic effluents, with many organic and inorganic contaminants in suspended and dissolved states. The turbidity restricts light penetration and oxygenation in the aquatic environment [188]. Based on the operational procedures employed and the type of oil processed, concentration and properties of organic contaminants differ, with COD values between 2,000 mg L^{-1} to 30,000 mg L^{-1} [189]. PEF exhibited promising results with EAOPs for elimination of 4-hydroxyphenylacetic acid in wastewater contaminated with olive oil mill effluents [190]. Total organic carbon (TOC) content reduced by 80% after 540 min, and biodegradability improved remarkably upon PEF treatment. In PEF, the rapid mineralization of photoactive by-products on exposure to UV-A light is most likely responsible for enhancing the efficiency of treatment procedure.

8.8.3 Pharmaceutical Waste

The effluents from pharmaceutical industry are composed of many contaminants, like synthetic or natural organic and inorganic compounds, catalysts, and many solvents [191]. The presence of a very high xenobiotic and active antibiotic concentrations makes the pharmaceutical wastewater very toxic with high COD and BOD and very low ratio of biodegradability [192]. Hence, traditional treatments strategies are not effective for degradation of pharmaceutical waste; this is evident from manifestation of antibiotic-resistant bacteria and genes [193] and some innovative treatment procedures need to be developed to effectively remove the pharmaceutical effluents. Trimethoprim (TMP) treatment in municipal wastewater used diverse techniques employing AOPs and EAOPs. The efficiency of TMP removal was maximum in PEF under UV-A irradiation, whereas UV-C exposure was inefficient [194]. Another study reported COD reduction in pharmaceutical waste under UV-A irradiation in PEF [195]. In another research, an effective cost-effective treatment was employed on combining UV photolysis and conductive diamond electrochemical oxidation [197].

8.8.4 Textile Industry

The textile waste contains different dyes, pesticides, salts, surfactants, polishing and coating agents, fabric softeners, and couplers. The composition varies from one lot to another [198], necessitating development of unique and innovative treatment techniques. In EAOPs, chloride ions can cause generation of toxic by-products of organochlorine degradation. Alves *et al.* reported a 65% decrease in chlorine containing by-products by employing photo-assisted EAOP, as compared to the ordinary electrochemical technique. According to another study, when UV irradiation is included in electrochemical system with $Ti/Ru_{0.3}Ti_{0.7}O_2$ as the anode, the organochlorine products are simultaneously formed and degraded, which enhances the TOC and COD removal by 13% and 30%, respectively, after 90 min of treatment time [199]. Likewise, in another study, a pilot-scale tubular flow reactor with 18-L capacity was designed using TiO_2/RuO_2 anode for treating textile waste by electrochemical and photoelectrochemical processes [200]. TOC and COD removal enhanced by an order magnitude upon irradiation with light. A solar photoreactor with a unipolar positioning of BDD/Ti anodes was created for treatment of dyes in wastewater [198]. The study reported a reduction of 83% COD after 15 min of treating the indigo dye effluent, along with an increase in the biodegradability index (BOD5/COD), which was elevated to above 0.4. Inclusion of Cu^{2+} and Fe^{2+} in sunlight-induced PEF, with an anode with BDD and a cathode with air-diffusion, facilitates rapid and complete disappearance of color along with induction of mineralization through promoting photo induced breakdown of Ferric complexes by solar irradiation and facilitating mineralization by oxidation of Cupric intermediates [201]. Electrochemically stable cellulose can block EAOP activity; hence, effective biodegradation of cellulose is required. Light-induced indirect electrochemical degradation performed on bio-treated textile waste, employing Ti/IrO_2-RuO_2-TiO_2 and Ti electrodes, under UV-visible light irradiation, effectively removed color by 98% and COD by 68%. Without photo-irradiation, COD removal was not reported, emphasizing the significance of oxychloride (^{-}OCl)

radicals generated upon UV-visible light irradiation [202]. Furthermore, in a research from Portugal, EC, PF, and activated sludge procedure were employed sequentially for effluent treatment, the COD being 1,200 mg L^{-1}) [203]. EC was followed by PF, resulting in almost 65% COD removal and appreciable increase in biodegradability, decreasing COD levels below 150 mg L^{-1} after completion of the activated sludge process.

8.8.5 Tannery Effluents

The tanning industry requires substantial water and releases almost 30 L to 35 L of wastewater in processing 1 kg of skin or hide [205]. The effluents contain acids, alkalies, solvents, chromium salts, sulfides, dyes, tannins, etc. [206]. The effluents have a high concentration of inorganic as well as organic contaminants. COD ranges between 1 g and 58 g L^{-1}, Cr^{6+} between 0.01 mg L^{-1} and 4.24 mg L^{-1}, Cr^{3+} between 0 g L^{-1} and 4.1 g L^{-1}, total dissolved solids between 22 g L^{-1} and 67 g L^{-1}, and sulfides between 0.025 g L^{-1} and 3.3 g L^{-1} [207]. Use of photo-assisted EAOPs for oxidative degradation of tannery waste has been studied extensively. An electrochemical cell of titanium cathode and Ti/IrO_2-RuO_2-TiO_2 anode was employed in a study and exposed to UV irradiation. The authors reported 100% sulfide, 92% COD, and 70% TOC removal, whereas, without UV exposure, the treatment was not very effective [208]. In another study, a BDD-based PEF process with UV-A irradiation was employed, which effectively mineralized organic compounds in wastewater [209]. A study employing photoperoxi electrocoagulation reported that the electrochemical and Fenton-based technique resulted in the removal of 80% COD, 95% color, 98% turbidity, 88% suspended solids, 96% total fixed solids, and 83% total volatile solids within 120 min of treatment time [207]. In another research, photolysis with exposure to vacuum UV (VUV) was used which reduced the limitation caused by turbidity and suspended particles by performing EC before photolysis. Hence, the unified effects of 20 min of EC followed by 40 min of photolysis eliminated 99% COD and 100% chromium with COD of 20,600 mg L^{-1} and total chromium of 30.11 mg L^{-1} [210].

8.9 Trends and Conclusion

Wastewater reclamation plays a crucial role in counteracting air, soil, and water pollution. Conventional treatment strategies are not giving promising results as the contaminants in water are highly toxic with a possibility of carcinogenic and mutagenic properties. Remediation of these contaminants is essential for sustaining the environment. Innovative strategies need to be developed to eliminate these toxic contaminants from water. Photoelectrochemical technique combined with exposure of light energy is an upcoming and promising procedure for removing contaminants from wastewater.

Photoelectrocatalysis with nanostructural components and their composites need further attention, and future studies need to stress on the role of nanocomposites in waste removal. Combination of several methodologies gives improved results. Fenton-based photoelectric processes are emerging as promising prospects for removal of recalcitrant contaminants.

Semiconductor photocatalysts have been studied extensively for removing organic contaminants from wastewater, but their widespread industrial application requires addressing a few concerns in order to increase their effectiveness and feasibility of application. Most of the commercially available photocatalysts are activated by UV irradiation but have low effectiveness on exposure to visible light. Development of a photocatalyst efficient in the visible spectrum and under natural sunlight has evoked immense interest in the scientific community working on reclamation of wastewater. In recent years, most investigations are focused on development of innovative composite photocatalysts to enhance their efficiency in removing contaminants. Challenges like high cost, complicated methodologies of synthesis, low stability, low efficiency, and lack of recyclability restrict their industrial application.

References

1. Veldkamp, T.I.E., Wada, Y., de Moel, H., Kummu, M., Eisner, S., Aerts, J.C.J.H., Ward, P.J., Changing mechanism of global water scarcity events: Impacts of socioeconomic changes and inter-annual hydro-climatic variability. *Global Environ. Change*, 32, 18–29, 2015, https://doi.org/10.1016/j.gloenvcha.2015.02.011.
2. Wada, Y., Flörke, M., Hanasaki, N., Eisner, S., Fischer, G., Tramberend, S., Satoh, Y., van Vliet, M.T.H., Yillia, P., Ringler, C., Burek, P., Wiberg, D., Modeling global water use for the 21st century: the Water Futures and Solutions (WFaS) initiative and its approaches. *Geosci. Model Dev.*, 9, 175–222, 2016, https://doi.org/10.5194/gmd-9-175-2016.
3. Sousa, J.C.G., Ribeiro, A.R., Barbosa, M.O., Pereira, M.F.R., Silva, A.M.T., A review on environmental monitoring of water organic pollutants identified by EU guidelines. *J. Hazard. Mater.*, 344, 146–162, 2018, https://doi.org/10.1016/j.jhazmat.2017.09.058.

4. Braslavsky, S.E., Braun, A.M., Cassano, A.E., Emeline, A.V., Litter, M.I., Palmisano, L., Parmon, V.N., Serpone, N., Glossary of terms used in photocatalysis and radiation catalysis (IUPAC Recommendations 2011). *Pure Appl. Chem.*, 83, 931–1014, 2011, https://doi.org/10.1351/PAC-REC-09-09-36.
5. Dehghani, M.H., Omrani, G.A., Karri, R.R., Solid Waste—Sources, Toxicity, and Their Consequences to Human Health, in: *Soft Computing Techniques in Solid Waste and Wastewater Management*, pp. 205–213, Elsevier: Amsterdam, Netherlands, 2021, https://doi.org/10.1016/B978-0-12-824463-0.00013-6.
6. Karri, R.R., Ravindran, G., Dehghani, M.H., Wastewater—Sources, Toxicity, and Their Consequences to Human Health, in: *Soft Computing Techniques in Solid Waste and Wastewater Management*, pp. 3–33, Elsevier: Amsterdam, Netherlands, 2021, https://doi.org/10.1016/B978-0-12-824463-0.00001-X.
7. Holkar, C.R., Jadhav, A.J., Pinjari, D.V., Mahamuni, N.M., Pandit, A.B., A critical review on textile wastewater treatments: Possible approaches. *J. Environ. Manag.*, 182, 351–366, 2016, https://doi.org/10.1016/j.jenvman.2016.07.090.
8. Feng, C., Yu, L., Xiao, Y., An, C., Environmental Behavior and Effects of Pollutants in Water. *J. Chem.*, 2020, 1–2, 2020, https://doi.org/10.1155/2020/2639389.
9. Orimolade, B.O. and Arotiba, O.A., Bismuth vanadate in photoelectrocatalytic water treatment systems for the degradation of organics: A review on recent trends. *J. Electroanal. Chem.*, 878, 114724, 2020, https://doi.org/10.1016/j.jelechem.2020.114724.
10. Mandeep, and Shukla, P., Microbial Nanotechnology for Bioremediation of Industrial Wastewater. *Front. Microbiol.*, 11, 590631, 2020, https://doi.org/10.3389/fmicb.2020.590631.
11. Arslan, M., Imran, A., Khan, Q.M., Afzal, M., Plant–bacteria partnerships for the remediation of persistent organic pollutants. *Environ. Sci. Pollut. Res.*, 24, 4322–4336, 2017, https://doi.org/10.1007/s11356-015-4935-3.
12. Khan, F.S.A., Mubarak, N.M., Khalid, M., Tan, Y.H., Abdullah, E.C., Rahman, M.E., Karri, R.R., A comprehensive review on micropollutants removal using carbon nanotubes-based adsorbents and membranes. *J. Environ. Chem. Eng.*, 9, 106647, 2021, https://doi.org/10.1016/j.jece.2021.106647.
13. Khan, F.S.A., Mubarak, N.M., Tan, Y.H., Khalid, M., Karri, R.R., Walvekar, R., Abdullah, E.C., Nizamuddin, S., Mazari, S.A., A comprehensive review on magnetic carbon nanotubes and carbon nanotube-based buckypaper for removal of heavy metals and dyes. *J. Hazard. Mater.*, 413, 125375, 2021, https://doi.org/10.1016/j.jhazmat.2021.125375.
14. Mehmood, A., Khan, F.S.A., Mubarak, N.M., Tan, Y.H., Karri, R.R., Khalid, M., Walvekar, R., Abdullah, E.C., Nizamuddin, S., Mazari, S.A., Magnetic nanocomposites for sustainable water purification—a comprehensive review. *Environ. Sci. Pollut. Res.*, 28, 19563–19588, 2021, https://doi.org/10.1007/s11356-021-12589-3.

15. Radha, K.V. and Sirisha, K., Electrochemical Oxidation Processes, in: *Advanced Oxidation Processes for Waste Water Treatment*, pp. 359–373, Elsevier, 2018, https://doi.org/10.1016/B978-0-12-810499-6.00011-5.
16. Vithalani, P.C. and Bhatt, N.S., Photoelectrochemical Water Treatment of Sewage, in: *Industrial Wastewater Treatment*, T. Karchiyappan, R.R. Karri, M.H. Dehghani (Eds.), pp. 399–420, Springer International Publishing, Cham: Amsterdam, Netherlands, 2022, https://doi.org/10.1007/978-3-030-98202-7_16.
17. Sunada, K., Watanabe, T., Hashimoto, K., Studies on photokilling of bacteria on TiO2 thin film. *J. Photochem. Photobiol. A*, 156, 227–233, 2003, https://doi.org/10.1016/S1010-6030(02)00434-3.
18. Aarthi, T. and Madras, G., Photocatalytic reduction of metals in presence of combustion synthesized nano-TiO2. *Catal. Commun.*, 9, 630–634, 2008, https://doi.org/10.1016/j.catcom.2007.07.001.
19. Muneer, M., Theurich, J., Bahnemann, D., Titanium dioxide mediated photocatalytic degradation of 1,2-diethyl phthalate. *J. Photochem. Photobiol. A*, 143, 213–219, 2001, https://doi.org/10.1016/S1010-6030(01)00525-1.
20. Varma, K.S., Tayade, R.J., Shah, K.J., Joshi, P.A., Shukla, A.D., Gandhi, V.G., Photocatalytic degradation of pharmaceutical and pesticide compounds (PPCs) using doped TiO2 nanomaterials: A review. *Water-Energy Nexus*, 3, 46–61, 2020, https://doi.org/10.1016/j.wen.2020.03.008.
21. Fujishima, A. and Honda, K., Electrochemical Photolysis of Water at a Semiconductor Electrode. *Nature*, 238, 37–38, 1972, https://doi.org/10.1038/238037a0.
22. Basavarajappa, P.S., Patil, S.B., Ganganagappa, N., Reddy, K.R., Raghu, A.V., Reddy, Ch.V., Recent progress in metal-doped TiO2, non-metal doped/codoped TiO2 and TiO2 nanostructured hybrids for enhanced photocatalysis. *Int. J. Hydrogen Energy*, 45, 7764–7778, 2020, https://doi.org/10.1016/j.ijhydene.2019.07.241.
23. Hanaor, D.A.H. and Sorrell, C.C., Review of the anatase to rutile phase transformation. *J. Mater. Sci.*, 46, 855–874, 2011, https://doi.org/10.1007/s10853-010-5113-0.
24. Scanlon, D.O., Dunnill, C.W., Buckeridge, J., Shevlin, S.A., Logsdail, A.J., Woodley, S.M., Catlow, C.R.A., Powell, M.J., Palgrave, R.G., Parkin, I.P., Watson, G.W., Keal, T.W., Sherwood, P., Walsh, A., Sokol, A.A., Band alignment of rutile and anatase TiO2. *Nat. Mater.*, 12, 798–801, 2013, https://doi.org/10.1038/nmat3697.
25. Jia, C., Zhang, X., Matras-Postolek, K., Huang, B., Yang, P., Z-scheme reduced graphene oxide/TiO2-Bronze/W18O49 ternary heterostructure towards efficient full solar-spectrum photocatalysis. *Carbon*, 139, 415–426, 2018, https://doi.org/10.1016/j.carbon.2018.07.024.
26. Byrne, C., Nolan, M., Banerjee, S., John, H., Jose, S., Periyat, P., Pillai, S.C., Advances in the Development of Novel Photocatalysts for Detoxification, in: *Visible Light-Active Photocatalysis*, pp. 283–327,

Wiley-VCH Verlag GmbH & Co. KGaA, Weinheim, Germany, 2018, https://doi.org/10.1002/9783527808175.ch11.
27. Imtiaz, F., Rashid, J., Xu, M., Semiconductor Nanocomposites for Visible Light Photocatalysis of Water Pollutants, in: *Concepts of Semiconductor Photocatalysis*, M. Rahman, A. Khan, A. Asiri, I. Inamuddin (Eds.), IntechOpen: London, UK, 2019, https://doi.org/10.5772/intechopen.86542.
28. Djurišić, A.B., He, Y., Ng, A.M.C., Visible-light photocatalysts: Prospects and challenges. *APL Mater.*, 8, 030903, 2020, https://doi.org/10.1063/1.5140497.
29. Ruan, L., Wang, X., Wang, T., Ren, Z., Chen, Y., Zhao, R., Zhou, D., Fu, G., Li, S., Gao, L., Lu, Y., Wang, Z., Tian, H., Kong, X., Han, G., Surface Defect-Controlled Growth and High Photocatalytic H_2 Production Efficiency of Anatase TiO_2 Nanosheets. *ACS Appl. Mater. Interfaces*, 11, 37256–37262, 2019, https://doi.org/10.1021/acsami.9b11233.
30. Saravanan, R., Gracia, F., Stephen, A., Basic Principles, Mechanism, and Challenges of Photocatalysis, in: *Nanocomposites for Visible Light-Induced Photocatalysis*, M.M. Khan, D. Pradhan, Y. Sohn (Eds.), pp. 19–40, Springer International Publishing, Cham, 2017, https://doi.org/10.1007/978-3-319-62446-4_2.
31. Rehman, S., Ullah, R., Butt, A.M., Gohar, N.D., Strategies of making TiO2 and ZnO visible light active. *J. Hazard. Mater.*, 170, 560–569, 2009, https://doi.org/10.1016/j.jhazmat.2009.05.064.
32. Byrne, C., Subramanian, G., Pillai, S.C., Recent advances in photocatalysis for environmental applications. *J. Environ. Chem. Eng.*, 6, 3531–3555, 2018, https://doi.org/10.1016/j.jece.2017.07.080.
33. Chen, Q., Zhang, Y., Zhang, D., Yang, Y., Ag and N co-doped TiO2 nano-structured photocatalyst for printing and dyeing wastewater. *J. Water Process Eng.*, 16, 14–20, 2017, https://doi.org/10.1016/j.jwpe.2016.11.007.
34. Chen, X., Wu, Z., Liu, D., Gao, Z., Preparation of ZnO Photocatalyst for the Efficient and Rapid Photocatalytic Degradation of Azo Dyes. *Nanoscale Res. Lett.*, 12, 143, 2017, https://doi.org/10.1186/s11671-017-1904-4.
35. Xu, L., Yang, L., Johansson, E.M.J., Wang, Y., Jin, P., Photocatalytic activity and mechanism of bisphenol a removal over TiO2–x/rGO nanocomposite driven by visible light. *Chem. Eng. J.*, 350, 1043–1055, 2018, https://doi.org/10.1016/j.cej.2018.06.046.
36. Marien, C.B.D., Marchal, C., Koch, A., Robert, D., Drogui, P., Sol-gel synthesis of TiO2 nanoparticles: effect of Pluronic P123 on particle's morphology and photocatalytic degradation of paraquat. *Environ. Sci. Pollut. Res.*, 24, 12582–12588, 2017, https://doi.org/10.1007/s11356-016-7681-2.
37. Gao, X., Peng, W., Tang, G., Guo, Q., Luo, Y., Highly efficient and visible-light-driven BiOCl for photocatalytic degradation of carbamazepine. *J. Alloys Compd.*, 757, 455–465, 2018, https://doi.org/10.1016/j.jallcom.2018.05.081.
38. Sahu, R.S., Shih, Y., Chen, W.-L., New insights of metal free 2D graphitic carbon nitride for photocatalytic degradation of bisphenol A. *J. Hazard. Mater.*, 402, 123509, 2021, https://doi.org/10.1016/j.jhazmat.2020.123509.

39. Wongcharoen, S. and Panomsuwan, G., Easy synthesis of TiO2 hollow fibers using kapok as a biotemplate for photocatalytic degradation of the herbicide paraquat. *Mater. Lett.*, 228, 482–485, 2018, https://doi.org/10.1016/j.matlet.2018.06.089.
40. Hazaraimi, M.H., Goh, P.S., Lau, W.J., Ismail, A.F., Wu, Z., Subramaniam, M.N., Lim, J.W., Kanakaraju, D., The state-of-the-art development of photocatalysts for the degradation of persistent herbicides in wastewater. *Sci. Total Environ.*, 843, 156975, 2022, https://doi.org/10.1016/j.scitotenv.2022.156975.
41. Byrne, J., Dunlop, P., Hamilton, J., Fernández-Ibáñez, P., Polo-López, I., Sharma, P., Vennard, A., A Review of Heterogeneous Photocatalysis for Water and Surface Disinfection. *Molecules*, 20, 5574–5615, 2015, https://doi.org/10.3390/molecules20045574.
42. McMichael, S., Fernández-Ibáñez, P., Byrne, J.A., A Review of Photoelectrocatalytic Reactors for Water and Wastewater Treatment. *Water*, 13, 1198, 2021, https://doi.org/10.3390/w13091198.
43. Zhang, J. and Nosaka, Y., Generation of OH radicals and oxidation mechanism in photocatalysis of WO3 and BiVO4 powders. *J. Photochem. Photobiol. A*, 303–304, 53–58, 2015, https://doi.org/10.1016/j.jphotochem.2015.01.008.
44. Nosaka, Y. and Nosaka, A.Y., Generation and Detection of Reactive Oxygen Species in Photocatalysis. *Chem. Rev.*, 117, 11302–11336, 2017, https://doi.org/10.1021/acs.chemrev.7b00161.
45. Bockris, J.O. and Otagawa, T., The Electrocatalysis of Oxygen Evolution on Perovskites. *J. Electrochem. Soc*, 131, 290–302, 1984, https://doi.org/10.1149/1.2115565.
46. Bockris, J., Khan, S., Murphy, O., Szklarczyk, M., On models of photoelectrocatalysis. *Int. J. Hydrogen Energy*, 9, 243–244, 1984, https://doi.org/10.1016/0360-3199(84)90125-3.
47. Sirés, I. and Brillas, E., Remediation of water pollution caused by pharmaceutical residues based on electrochemical separation and degradation technologies: A review. *Environ. Int.*, 40, 212–229, 2012, https://doi.org/10.1016/j.envint.2011.07.012.
48. Meng, X., Zhang, Z., Li, X., Synergetic photoelectrocatalytic reactors for environmental remediation: A review. *J. Photochem. Photobiol. C*, 24, 83–101, 2015, https://doi.org/10.1016/j.jphotochemrev.2015.07.003.
49. Janotti, A. and Van de Walle, C.G., Fundamentals of zinc oxide as a semiconductor. *Rep. Prog. Phys.*, 72, 126501, 2009, https://doi.org/10.1088/0034-4885/72/12/126501.
50. Wang, Y., Wu, Q., Li, Y., Liu, L., Geng, Z., Li, Y., Chen, J., Bai, W., Jiang, G., Zhao, Z., Controlled fabrication of TiO2/C3N4 core–shell nanowire arrays: a visible-light-responsive and environmental-friendly electrode for photoelectrocatalytic degradation of bisphenol A. *J. Mater. Sci.*, 53, 11015–11026, 2018, https://doi.org/10.1007/s10853-018-2368-3.

51. Xu, Z., Li, X., Li, J., Wu, L., Zeng, Q., Zhou, Z., Effect of CoOOH loading on the photoelectrocatalytic performance of WO3 nanorod array film. *Appl. Surf. Sci.*, 284, 285–290, 2013, https://doi.org/10.1016/j.apsusc.2013.07.095.
52. Kim, S.-G., Dhandole, L.K., Seo, Y.-S., Chung, H.-S., Chae, W.-S., Cho, M., Jang, J.S., Active composite photocatalyst synthesized from inactive Rh & Sb doped TiO2 nanorods: Enhanced degradation of organic pollutants & antibacterial activity under visible light irradiation. *Appl. Catal. A*, 564, 43–55, 2018, https://doi.org/10.1016/j.apcata.2018.07.016.
53. Klingshirn, C., ZnO: Material, Physics and Applications. *ChemPhysChem*, 8, 782–803, 2007, https://doi.org/10.1002/cphc.200700002.
54. Yagi, M., Maruyama, S., Sone, K., Nagai, K., Norimatsu, T., Preparation and photoelectrocatalytic activity of a nano-structured WO3 platelet film. *J. Solid State Chem.*, 181, 175–182, 2008, https://doi.org/10.1016/j.jssc.2007.11.018.
55. Hepel, M. and Luo, J., Photoelectrochemical mineralization of textile diazo dye pollutants using nanocrystalline WO3 electrodes. *Electrochim. Acta*, 47, 729–740, 2001, https://doi.org/10.1016/S0013-4686(01)00753-8.
56. Mahadik, M., Shinde, S., Mohite, V., Kumbhar, S., Rajpure, K., Moholkar, A., Kim, J., Bhosale, C., Photoelectrocatalytic oxidation of Rhodamine B with sprayed α-Fe_2O_3 photocatalyst. *Mat. Express*, 3, 247–255, 2013, https://doi.org/10.1166/mex.2013.1120.
57. Zhang, M., Pu, W., Pan, S., Okoth, O.K., Yang, C., Zhang, J., Photoelectrocatalytic activity of liquid phase deposited α-Fe2O3 films under visible light illumination. *J. Alloys Compd.*, 648, 719–725, 2015, https://doi.org/10.1016/j.jallcom.2015.07.026.
58. Yu, S., Xi, M., Han, K., Wang, Z., Yang, W., Zhu, H., Preparation and photoelectrocatalytic properties of polyaniline/layered manganese oxide self-assembled film. *Thin Solid Films*, 519, 357–361, 2010, https://doi.org/10.1016/j.tsf.2010.07.105.
59. Fan, C.M., Hua, B., Wang, Y., Liang, Z.H., Hao, X.G., Liu, S.B., Sun, Y.P., Preparation of Ti/SnO2–Sb2O4 photoanode by electrodeposition and dip coating for PEC oxidations. *Desalination*, 249, 736–741, 2009, https://doi.org/10.1016/j.desal.2009.01.035.
60. Wang, Y., Fan, C., Hua, B., Liang, Z., Sun, Y., Photoelectrocatalytic activity of two antimony doped SnO2 films for oxidation of phenol pollutants. *Trans. Nonferrous Met. Soc China*, 19, 778–783, 2009, https://doi.org/10.1016/S1003-6326(08)60349-0.
61. Florencio, J., Pacheco, M.J., Lopes, A., Ciriaco, L., Application of Ti/Pt/beta-PbO2 Anodes in the Degradation of DR80 Azo Dye. *Port. Electrochim. Acta*, 31, 257–264, 2013, https://doi.org/10.4152/pea.201305257.
62. da Silva, M.R., Lucilha, A.C., Afonso, R., Dall'Antonia, L.H., de Andrade Scalvi, L.V., Photoelectrochemical properties of FTO/m-BiVO4 electrode in different electrolytes solutions under visible light irradiation. *Ionics.*, 20, 105–113, 2014, https://doi.org/10.1007/s11581-013-0967-1.

63. Pan, C. and Zhu, Y., New Type of $BiPO_4$ Oxy-Acid Salt Photocatalyst with High Photocatalytic Activity on Degradation of Dye. *Environ. Sci. Technol.*, 44, 5570–5574, 2010, https://doi.org/10.1021/es101223n.
64. Kerkez, Ö. and Boz, İ., Photo(electro)catalytic Activity of Cu2+-Modified TiO2 Nanorod Array Thin Films under Visible Light Irradiation. *J. Phys. Chem. Solids*, 75, 611–618, 2014, https://doi.org/10.1016/j.jpcs.2013.12.019.
65. Zhang, Z., Yu, Y., Wang, P., Hierarchical Top-Porous/Bottom-Tubular TiO_2 Nanostructures Decorated with Pd Nanoparticles for Efficient Photoelectrocatalytic Decomposition of Synergistic Pollutants. *ACS Appl. Mater. Interfaces*, 4, 990–996, 2012, https://doi.org/10.1021/am201630s.
66. Javaid, R. and Qazi, U.Y., Catalytic Oxidation Process for the Degradation of Synthetic Dyes: An Overview. *Int. J. Environ. Res. Public Health*, 16, 2066, 2019, https://doi.org/10.3390/ijerph16112066.
67. Susan, E.F. and Elizabeth, A.D., Sensitization to dyes in textiles and other consumer products. *J. Toxicol.: Cutaneous Ocul. Toxicol.*, 7, 195–222, 1988, https://doi.org/10.3109/15569528809052329.
68. Svedman, C., Engfeldt, M., Malinauskiene, L., Textile Contact Dermatitis: How Fabrics Can Induce Dermatitis. *Curr. Treat. Options Allergy*, 6, 103–111, 2019, https://doi.org/10.1007/s40521-019-0197-5.
69. Jamee, R. and Siddique, R., Biodegradation of synthetic dyes of textile effluent by microorganisms: an environmentally and economically sustainable approach. *Eur. J. Microbiol. Immunol.*, 9, 114–118, 2019, https://doi.org/10.1556/1886.2019.00018.
70. Mancuso, A. and Iervolino, G., Synthesis and Application of Innovative and Environmentally Friendly Photocatalysts: A Review. *Catalysts*, 12, 1074, 2022, https://doi.org/10.3390/catal12101074.
71. Ahmed, S.F., Mofijur, M., Nuzhat, S., Chowdhury, A.T., Rafa, N., Uddin, Md.A., Inayat, A., Mahlia, T.M.I., Ong, H.C., Chia, W.Y., Show, P.L., Recent developments in physical, biological, chemical, and hybrid treatment techniques for removing emerging contaminants from wastewater. *J. Hazard. Mater.*, 416, 125912, 2021, https://doi.org/10.1016/j.jhazmat.2021.125912.
72. Kasonga, T.K., Coetzee, M.A.A., Kamika, I., Ngole-Jeme, V.M., Benteke Momba, M.N., Endocrine-disruptive chemicals as contaminants of emerging concern in wastewater and surface water: A review. *J. Environ. Manag.*, 277, 111485, 2021, https://doi.org/10.1016/j.jenvman.2020.111485.
73. Pal, A., Gin, K.Y.-H., Lin, A.Y.-C., Reinhard, M., Impacts of emerging organic contaminants on freshwater resources: Review of recent occurrences, sources, fate and effects. *Sci. Total Environ.*, 408, 6062–6069, 2010, https://doi.org/10.1016/j.scitotenv.2010.09.026.
74. Akhbarizadeh, R., Dobaradaran, S., Schmidt, T.C., Nabipour, I., Spitz, J., Worldwide bottled water occurrence of emerging contaminants: A review of the recent scientific literature. *J. Hazard. Mater.*, 392, 122271, 2020, https://doi.org/10.1016/j.jhazmat.2020.122271.

75. Jin, S., Guerrero-Juarez, C.F., Zhang, L., Chang, I., Ramos, R., Kuan, C.-H., Myung, P., Plikus, M.V., Nie, Q., Inference and analysis of cell-cell communication using CellChat. *Nat. Commun.*, 12, 1088, 2021, https://doi.org/10.1038/s41467-021-21246-9.
76. Jiang, J.-J., Lee, C.-L., Fang, M.-D., Tu, B.-W., Liang, Y.-J., Impacts of Emerging Contaminants on Surrounding Aquatic Environment from a Youth Festival. *Environ. Sci. Technol.*, 49, 792–799, 2015, https://doi.org/10.1021/es503944e.
77. Saidulu, D., Gupta, B., Gupta, A.K., Ghosal, P.S., A review on occurrences, eco-toxic effects, and remediation of emerging contaminants from wastewater: Special emphasis on biological treatment based hybrid systems. *J. Environ. Chem. Eng.*, 9, 105282, 2021, https://doi.org/10.1016/j.jece.2021.105282.
78. Zhang, B., He, X., Yu, C., Liu, G., Ma, D., Cui, C., Yan, Q., Zhang, Y., Zhang, G., Ma, J., Xin, Y., Degradation of tetracycline hydrochloride by ultrafine TiO2 nanoparticles modified g-C3N4 heterojunction photocatalyst: Influencing factors, products and mechanism insight. *Chin. Chem. Lett.*, 33, 1337–1342, 2022, https://doi.org/10.1016/j.cclet.2021.08.008.
79. Prieto-Rodriguez, L., Miralles-Cuevas, S., Oller, I., Agüera, A., Puma, G.L., Malato, S., Treatment of emerging contaminants in wastewater treatment plants (WWTP) effluents by solar photocatalysis using low TiO2 concentrations. *J. Hazard. Mater.*, 211–212, 131–137, 2012, https://doi.org/10.1016/j.jhazmat.2011.09.008.
80. López-Gámez, G., Elez-Martínez, P., Martín-Belloso, O., Soliva-Fortuny, R., Pulsed electric fields affect endogenous enzyme activities, respiration and biosynthesis of phenolic compounds in carrots. *Postharvest Biol. Technol.*, 168, 111284, 2020, https://doi.org/10.1016/j.postharvbio.2020.111284.
81. Hays, M.D., Fine, P.M., Geron, C.D., Kleeman, M.J., Gullett, B.K., Open burning of agricultural biomass: Physical and chemical properties of particle-phase emissions. *Atmos. Environ.*, 39, 6747–6764, 2005, https://doi.org/10.1016/j.atmosenv.2005.07.072.
82. Batura, I.I., Gogotov, A.F., Cherepanov, V.I., Baranov, O.I., Levchuk, A.A., Parilova, M.V., Utilizing the phenol byproducts of coke production: 3. Phenols as coinhibitors of thermopolymerization during styrene production. *Coke Chem.*, 52, 23–27, 2009, https://doi.org/10.3103/S1068364X09010074.
83. Villegas, L.G.C., Mashhadi, N., Chen, M., Mukherjee, D., Taylor, K.E., Biswas, N., A Short Review of Techniques for Phenol Removal from Wastewater. *Curr. Pollut. Rep.*, 2, 157–167, 2016, https://doi.org/10.1007/s40726-016-0035-3.
84. Pedroza, A.M., Mosqueda, R., Alonso-Vante, N., Rodríguez-Vázquez, R., Sequential treatment via Trametes versicolor and UV/TiO2/RuxSey to reduce contaminants in waste water resulting from the bleaching process during paper production. *Chemosphere*, 67, 793–801, 2007, https://doi.org/10.1016/j.chemosphere.2006.10.015.
85. She, Z., Gao, M., Jin, C., Chen, Y., Yu, J., Toxicity and biodegradation of 2,4-dinitrophenol and 3-nitrophenol in anaerobic systems. *Process Biochem.*, 40, 3017–3024, 2005, https://doi.org/10.1016/j.procbio.2005.02.007.

86. Gopi, P.K., Srinithi, S., Chen, S.-M., Ravikumar, C.H., Designing of cerium-doped bismuth vanadate nanorods/functionalized-MWCNT nanocomposite for the high toxicity of 4-cyanophenol herbicide detection in human urine sample. *Colloids Surf. A*, 639, 128371, 2022, https://doi.org/10.1016/j.colsurfa.2022.128371.
87. Tyler, C.R., Jobling, S., Sumpter, J.P., Endocrine Disruption in Wildlife: A Critical Review of the Evidence. *Crit. Rev. Toxicol.*, 28, 319–361, 1998, https://doi.org/10.1080/10408449891344236.
88. Karci, A., Arslan-Alaton, I., Olmez-Hanci, T., Bekbölet, M., Transformation of 2,4-dichlorophenol by H2O2/UV-C, Fenton and photo-Fenton processes: Oxidation products and toxicity evolution. *J. Photochem. Photobiol., A*, 230, 65–73, 2012, https://doi.org/10.1016/j.jphotochem.2012.01.003.
89. Niu, Y., Xing, M., Zhang, J., Tian, B., Visible light activated sulfur and iron co-doped TiO2 photocatalyst for the photocatalytic degradation of phenol. *Catal. Today*, 201, 159–166, 2013, https://doi.org/10.1016/j.cattod.2012.04.035.
90. Xu, J., Wang, F., Liu, W., Cao, W., Nanocrystalline N-Doped TiO 2 Powders: Mild Hydrothermal Synthesis and Photocatalytic Degradation of Phenol under Visible Light Irradiation. *Int. J. Photoenergy*, 2013, 1–7, 2013, https://doi.org/10.1155/2013/616139.
91. Vaiano, V., Lara, M.A., Iervolino, G., Matarangolo, M., Navio, J.A., Hidalgo, M.C., Photocatalytic H2 production from glycerol aqueous solutions over fluorinated Pt-TiO2 with high {001} facet exposure. *J. Photochem. Photobiol., A*, 365, 52–59, 2018, https://doi.org/10.1016/j.jphotochem.2018.07.032.
92. Iervolino, G., Vaiano, V., Sannino, D., Puga, F., Navío, J.A., Hidalgo, M.C., LaFeO3 Modified with Ni for Hydrogen Evolution via Photocatalytic Glucose Reforming in Liquid Phase. *Catalysts*, 11, 1558, 2021, https://doi.org/10.3390/catal11121558.
93. Qian, W., Greaney, P.A., Fowler, S., Chiu, S.-K., Goforth, A.M., Jiao, J., Low-Temperature Nitrogen Doping in Ammonia Solution for Production of N-Doped TiO_2-Hybridized Graphene as a Highly Efficient Photocatalyst for Water Treatment. *ACS Sustain. Chem. Eng.*, 2, 1802–1810, 2014, https://doi.org/10.1021/sc5001176.
94. Liu, X., Xing, Z., Zhang, Y., Li, Z., Wu, X., Tan, S., Yu, X., Zhu, Q., Zhou, W., Fabrication of 3D flower-like black N-TiO2-x@MoS2 for unprecedented-high visible-light-driven photocatalytic performance. *Appl. Catal. B: Environ.*, 201, 119–127, 2017, https://doi.org/10.1016/j.apcatb.2016.08.031.
95. Nadeem, M.A. and Idriss, H., Photo-thermal reactions of ethanol over Ag/TiO_2 catalysts. The role of silver plasmon resonance in the reaction kinetics. *Chem. Commun.*, 54, 5197–5200, 2018, https://doi.org/10.1039/C8CC01814F.
96. Wang, W., Zhao, H., Pan, P., Xue, K., Zhang, Z., Duan, J., Enhanced the photocatalytic activity of B–C–N–TiO2 under visible light:Synergistic effect of element doping and Z-scheme interface heterojunction constructed with Ag

nanoparticles. *Ceram. Int.*, 47, 6094–6104, 2021, https://doi.org/10.1016/j.ceramint.2020.10.188.

97. Soni, V., Raizada, P., Kumar, A., Hasija, V., Singal, S., Singh, P., Hosseini-Bandegharaei, A., Thakur, V.K., Nguyen, V.-H., Indium sulfide-based photocatalysts for hydrogen production and water cleaning: a review. *Environ. Chem. Lett.*, 19, 1065–1095, 2021, https://doi.org/10.1007/s10311-020-01148-w.
98. He, J., Kumar, A., Khan, M., Lo, I.M.C., Critical review of photocatalytic disinfection of bacteria: from noble metals- and carbon nanomaterials-TiO2 composites to challenges of water characteristics and strategic solutions. *Sci. Total Environ.*, 758, 143953, 2021, https://doi.org/10.1016/j.scitotenv.2020.143953.
99. Jeyalakshmi, V., Mahalakshmy, R., Krishnamurthy, K.R., Viswanathan, B., Photocatalytic Reduction of Carbon Dioxide by Water: A Step towards Sustainable Fuels and Chemicals. *Mater. Sci. Forum*, 734, 1–62, 2012, https://doi.org/10.4028/www.scientific.net/MSF.734.1.
100. Basu, A., Misra, A.J., Behera, M., Behera, S.K., Nayak, A.K., Dhal, N.K., Mishra, A., Satpathy, B.K., Lundborg, C.S., Tripathy, S.K., Photocatalytic disinfection of extended-spectrum beta-lactamase producing Escherichia coli using Alumina/ZnO heterostructures. *J. Environ. Chem. Eng.*, 9, 106334, 2021, https://doi.org/10.1016/j.jece.2021.106334.
101. Sethi, D. and Sakthivel, R., ZnO/TiO2 composites for photocatalytic inactivation of Escherichia coli. *J. Photochem. Photobiol. B*, 168, 117–123, 2017, https://doi.org/10.1016/j.jphotobiol.2017.02.005.
102. Mangayayam, M., Kiwi, J., Giannakis, S., Pulgarin, C., Zivkovic, I., Magrez, A., Rtimi, S., FeOx magnetization enhancing E. coli inactivation by orders of magnitude on Ag-TiO2 nanotubes under sunlight. *Appl. Catal. B: Environ.*, 202, 438–445, 2017, https://doi.org/10.1016/j.apcatb.2016.09.064.
103. Yan, H., Liu, L., Wang, R., Zhu, W., Ren, X., Luo, L., Zhang, X., Luo, S., Ai, X., Wang, J., Binary composite MoS2/TiO2 nanotube arrays as a recyclable and efficient photocatalyst for solar water disinfection. *Chem. Eng. J.*, 401, 126052, 2020, https://doi.org/10.1016/j.cej.2020.126052.
104. Lofrano, G., Ubaldi, F., Albarano, L., Carotenuto, M., Vaiano, V., Valeriani, F., Libralato, G., Gianfranceschi, G., Fratoddi, I., Meric, S., Guida, M., Romano Spica, V., Antimicrobial Effectiveness of Innovative Photocatalysts: A Review. *Nanomaterials*, 12, 2831, 2022, https://doi.org/10.3390/nano12162831.
105. Wang, J., Zhuang, H., Hinton, A., Bowker, B., Zhang, J., Photocatalytic disinfection of spoilage bacteria Pseudomonas fluorescens and Macrococcus caseolyticus by nano-TiO2. *LWT - Food Sci. Technol.*, 59, 1009–1017, 2014, https://doi.org/10.1016/j.lwt.2014.06.062.
106. Zhang, C., Zhang, M., Li, Y., Shuai, D., Visible-light-driven photocatalytic disinfection of human adenovirus by a novel heterostructure of oxygen-doped graphitic carbon nitride and hydrothermal carbonation carbon. *Appl. Catal. B: Environ.*, 248, 11–21, 2019, https://doi.org/10.1016/j.apcatb.2019.02.009.

107. Baniamerian, H., Safavi, M., Alvarado-Morales, M., Tsapekos, P., Angelidaki, I., Shokrollahzadeh, S., Photocatalytic inactivation of Vibrio fischeri using Fe2O3-TiO2-based nanoparticles. *Environ. Res.*, 166, 497–506, 2018, https://doi.org/10.1016/j.envres.2018.06.011.
108. Liang, J., Deng, J., Li, M., Tong, M., Bactericidal activity and mechanism of AgI/AgBr/BiOBr0.75I0.25 under visible light irradiation. *Colloids Surf. B*, 138, 102–109, 2016, https://doi.org/10.1016/j.colsurfb.2015.11.060.
109. Liang, J., Deng, J., Li, M., Xu, T., Tong, M., Bactericidal activity and mechanism of Ti-doped BiOI microspheres under visible light irradiation. *Colloids Surf. B*, 147, 307–314, 2016, https://doi.org/10.1016/j.colsurfb.2016.08.016.
110. Feng, T., Liang, J., Ma, Z., Li, M., Tong, M., Bactericidal activity and mechanisms of BiOBr-AgBr under both dark and visible light irradiation conditions. *Colloids Surf. B*, 167, 275–283, 2018, https://doi.org/10.1016/j.colsurfb.2018.04.022.
111. Li, Y., Li, Y., Ma, S., Wang, P., Hou, Q., Han, J., Zhan, S., Efficient water disinfection with Ag 2 WO 4 -doped mesoporous g-C 3 N 4 under visible light. *J. Hazard. Mater.*, 338, 33–46, 2017, https://doi.org/10.1016/j.jhazmat.2017.05.011.
112. Song, J., Yu, J., Sun, G., Si, Y., Ding, B., Visible-light-driven, hierarchically heterostructured, and flexible silver/bismuth oxyiodide/titania nanofibrous membranes for highly efficient water disinfection. *J. Colloid Interface Sci.*, 555, 636–646, 2019, https://doi.org/10.1016/j.jcis.2019.08.017.
113. Wu, X., Cao, L., Song, J., Si, Y., Yu, J., Ding, B., Thorn-like flexible Ag2C2O4/TiO2 nanofibers as hierarchical heterojunction photocatalysts for efficient visible-light-driven bacteria-killing. *J. Colloid Interface Sci.*, 560, 681–689, 2020, https://doi.org/10.1016/j.jcis.2019.10.119.
114. Zhang, G., Savateev, A., Zhao, Y., Li, L., Antonietti, M., Advancing the n → π* electron transition of carbon nitride nanotubes for H_2 photosynthesis. *J. Mater. Chem.A*, 5, 12723–12728, 2017, https://doi.org/10.1039/C7TA03777E.
115. Zhang, C., Li, Y., Shuai, D., Zhang, W., Niu, L., Wang, L., Zhang, H., Visible-light-driven, water-surface-floating antimicrobials developed from graphitic carbon nitride and expanded perlite for water disinfection. *Chemosphere*, 208, 84–92, 2018, https://doi.org/10.1016/j.chemosphere.2018.05.163.
116. Shanmugam, V., Sanjeevamuthu, S., Jeyaperumal, K.S., Vairamuthu, R., Fabrication of heterostructured vanadium modified g-C3N4/TiO2 hybrid photocatalyst for improved photocatalytic performance under visible light exposure and antibacterial activities. *J. Ind. Eng. Chem.*, 76, 318–332, 2019, https://doi.org/10.1016/j.jiec.2019.03.056.
117. Deng, J., Liang, J., Li, M., Tong, M., Enhanced visible-light-driven photocatalytic bacteria disinfection by g-C 3 N 4 -AgBr. *Colloids Surf. B*, 152, 49–57, 2017, https://doi.org/10.1016/j.colsurfb.2017.01.003.
118. Sun, L., Du, T., Hu, C., Chen, J., Lu, J., Lu, Z., Han, H., Antibacterial Activity of Graphene Oxide/g-C_3N_4 Composite through Photocatalytic Disinfection

under Visible Light. *ACS Sustain. Chem. Eng.*, 5, 8693–8701, 2017, https://doi.org/10.1021/acssuschemeng.7b01431.
119. Wang, G., Chang, J., Tang, W., Xie, W., Ang, Y.S., 2D materials and heterostructures for photocatalytic water-splitting: a theoretical perspective. *J. Phys. D: Appl. Phys.*, 55, 293002, 2022, https://doi.org/10.1088/1361-6463/ac5771.
120. Xu, L., Zhao, J., Liu, Z., Wang, Z., Yu, K., Xing, B., Cleavage and transformation inhibition of extracellular antibiotic resistance genes by graphene oxides with different lateral sizes. *Sci. Total Environ.*, 695, 133932, 2019, https://doi.org/10.1016/j.scitotenv.2019.133932.
121. O'Dowd, K., Nair, K.M., Pillai, S.C., Photocatalytic degradation of antibiotic-resistant genes and bacteria using 2D nanomaterials: What is known and what are the challenges? *Curr. Opin. Green Sustain. Chem.*, 30, 100471, 2021, https://doi.org/10.1016/j.cogsc.2021.100471.
122. Shi, H., Wang, C., Zhao, Y., Liu, E., Fan, J., Ji, Z., Highly efficient visible light driven photocatalytic inactivation of E. coli with Ag QDs decorated Z-scheme Bi2S3/SnIn4S8 composite. *Appl. Catal. B: Environ.*, 254, 403–413, 2019, https://doi.org/10.1016/j.apcatb.2019.05.020.
123. Xia, D., An, T., Li, G., Wang, W., Zhao, H., Wong, P.K., Synergistic photocatalytic inactivation mechanisms of bacteria by graphene sheets grafted plasmonic Ag AgX (X = Cl, Br, I) composite photocatalyst under visible light irradiation. *Water. Res.*, 99, 149–161, 2016, https://doi.org/10.1016/j.watres.2016.04.055.
124. Zinicovscaia, I. and Cepoi, L. (Eds.), *Cyanobacteria for Bioremediation of Wastewaters*, Springer International Publishing, Cham, 2016, https://doi.org/10.1007/978-3-319-26751-7.
125. Trivedi, N.S. and Mandavgane, S.A., Fundamentals of 2, 4 Dichloropheno xyacetic Acid Removal from Aqueous Solutions. *Sep. Purif. Rev.*, 47, 337–354, 2018, https://doi.org/10.1080/15422119.2018.1450765.
126. Saleh, I.A., Zouari, N., Al-Ghouti, M.A., Removal of pesticides from water and wastewater: Chemical, physical and biological treatment approaches. *Environ. Technol. Innovation.*, 19, 101026, 2020, https://doi.org/10.1016/j.eti.2020.101026.
127. Chiam, C. and Sarbatly, R., Purification of Aquacultural Water: Conventional and New Membrane-based Techniques. *Sep. Purif. Rev.*, 40, 126–160, 2011, https://doi.org/10.1080/15422119.2010.549766.
128. Zhang, J., Weston, G., Yang, X., Gray, S., Duke, M., Removal of herbicide 2-methyl-4-chlorophenoxyacetic acid (MCPA) from saline industrial wastewater by reverse osmosis and nanofiltration. *Desalination*, 496, 114691, 2020, https://doi.org/10.1016/j.desal.2020.114691.
129. Igunnu, E.T. and Chen, G.Z., Produced water treatment technologies. *Int. J. Low-Carbon Technol.*, 9, 157–177, 2014, https://doi.org/10.1093/ijlct/cts049.
130. Subramaniam, M.N., Goh, P.-S., Lau, W.-J., Ng, B.-C., Ismail, A.F., Development of nanomaterial-based photocatalytic membrane for organic

pollutants removal, in: *Advanced Nanomaterials for Membrane Synthesis and Its Applications*, pp. 45–67, Elsevier: Cambridge, Massachusetts (US), 2019, https://doi.org/10.1016/B978-0-12-814503-6.00003-3.
131. Abdelhaleem, A. and Chu, W., Photodegradation of 4-chlorophenoxyacetic acid under visible LED activated N-doped TiO 2 and the mechanism of stepwise rate increment of the reused catalyst. *J. Hazard. Mater.*, 338, 491–501, 2017, https://doi.org/10.1016/j.jhazmat.2017.05.056.
132. Nasirian, M., Lin, Y.P., Bustillo-Lecompte, C.F., Mehrvar, M., Enhancement of photocatalytic activity of titanium dioxide using non-metal doping methods under visible light: a review. *Int. J. Environ. Sci. Technol.*, 15, 2009–2032, 2018, https://doi.org/10.1007/s13762-017-1618-2.
133. Suwannaruang, T. and Wantala, K., Single-step uncalcined N-TiO 2 synthesis, characterizations and its applications on alachlor photocatalytic degradations. *Appl. Surf. Sci.*, 380, 257–267, 2016, https://doi.org/10.1016/j.apsusc.2016.01.117.
134. Hosseini, N. and Toosi, M.R., Combined adsorption process and photocatalytic degradation of some commercial herbicides over N-doped TiO_2 particles supported on recyclable magnetic hexagonal mesoporous silica. *Sep. Sci. Technol.*, 54, 1697–1709, 2019, https://doi.org/10.1080/01496395.2018.1539105.
135. Wang, W.-K., Chen, J.-J., Gao, M., Huang, Y.-X., Zhang, X., Yu, H.-Q., Photocatalytic degradation of atrazine by boron-doped TiO2 with a tunable rutile/anatase ratio. *Appl. Catal. B: Environ.*, 195, 69–76, 2016, https://doi.org/10.1016/j.apcatb.2016.05.009.
136. Salah, N., Hameed, A., Aslam, M., Abdel-wahab, M.Sh., Babkair, S.S., Bahabri, F.S., Flow controlled fabrication of N doped ZnO thin films and estimation of their performance for sunlight photocatalytic decontamination of water. *Chem. Eng. J.*, 291, 115–127, 2016, https://doi.org/10.1016/j.cej.2016.01.111.
137. Shao, J., Sheng, W., Wang, M., Li, S., Chen, J., Zhang, Y., Cao, S., *In situ* synthesis of carbon-doped TiO 2 single-crystal nanorods with a remarkably photocatalytic efficiency. *Appl. Catal. B: Environ.*, 209, 311–319, 2017, https://doi.org/10.1016/j.apcatb.2017.03.008.
138. Sudrajat, H., Template-free, simple fabrication of C/N-doped Bi2O3 nanospheres with appreciable photocatalytic activity under visible light. *Superlattices Microstruct.*, 109, 229–239, 2017, https://doi.org/10.1016/j.spmi.2017.05.010.
139. Shamsedini, N., Dehghani, M., Nasseri, S., Baghapour, M.A., Photocatalytic degradation of atrazine herbicide with Illuminated Fe+3-TiO2 Nanoparticles. *J. Environ. Health Sci. Eng.*, 15, 7, 2017, https://doi.org/10.1186/s40201-017-0270-6.
140. Mendiola-Alvarez, S.Y., Guzmán-Mar, J.L., Turnes-Palomino, G., Maya-Alejandro, F., Hernández-Ramírez, A., Hinojosa-Reyes, L., UV and visible activation of Cr(III)-doped TiO2 catalyst prepared by a microwave-assisted

sol–gel method during MCPA degradation. *Environ. Sci. Pollut. Res.*, 24, 12673–12682, 2017, https://doi.org/10.1007/s11356-016-8034-x.

141. Gholami, M., Jonidi-Jafari, A., Farzadkia, M., Esrafili, A., Godini, K., Shirzad-Siboni, M., Photocatalytic removal of bentazon by copper doped zinc oxide nanorods: Reaction pathways and toxicity studies. *J. Environ. Manag.*, 294, 112962, 2021, https://doi.org/10.1016/j.jenvman.2021.112962.
142. Zhang, F., Wang, X., Liu, H., Liu, C., Wan, Y., Long, Y., Cai, Z., Recent Advances and Applications of Semiconductor Photocatalytic Technology. *Appl. Sci.*, 9, 2489, 2019, https://doi.org/10.3390/app9122489.
143. Majumdar, A. and Pal, A., Recent advancements in visible-light-assisted photocatalytic removal of aqueous pharmaceutical pollutants. *Clean Technol. Environ. Policy*, 22, 11–42, 2020, https://doi.org/10.1007/s10098-019-01766-1.
144. Saidani, M.A., Fkiri, A., Smiri, L.-S., Facile Synthesis of Ag/ZnO Photocatalysts on the Degradation of Diuron Herbicide Under Simulated Solar Light and the Investigation of Its Antibacterial Activity for Waste-Water Treatment. *J. Inorg. Organomet. Polym.*, 29, 710–720, 2019, https://doi.org/10.1007/s10904-018-1044-z.
145. Mkhalid, I.A., Photocatalytic degradation of herbicides under visible light using Pd-WO3 nanorods. *Ceram. Int.*, 42, 15975–15980, 2016, https://doi.org/10.1016/j.ceramint.2016.07.100.
146. González, A.E., Asomoza, M., Solís, S., Sánchez, M.A.G., Cipagauta-Díaz, S., Enhanced photocatalytic degradation of the herbicide 2,4-dichlorophenoxyacetic acid by Pt/TiO2–SiO2 nanocomposites. *React. Kinet. Mech. Catal.*, 131, 489–503, 2020, https://doi.org/10.1007/s11144-020-01865-x.
147. Lam, S.-M., Jaffari, Z.H., Sin, J.-C., Zeng, H., Lin, H., Li, H., Mohamed, A.R., Insight into the influence of noble metal decorated on BiFeO3 for 2,4-dichlorophenol and real herbicide wastewater treatment under visible light. *Colloids Surf. A.*, 614, 126138, 2021, https://doi.org/10.1016/j.colsurfa.2021.126138.
148. Brillas, E., A review on the photoelectro-Fenton process as efficient electrochemical advanced oxidation for wastewater remediation. Treatment with UV light, sunlight, and coupling with conventional and other photo-assisted advanced technologies. *Chemosphere*, 250, 126198, 2020, https://doi.org/10.1016/j.chemosphere.2020.126198.
149. Sirés, I., Brillas, E., Oturan, M.A., Rodrigo, M.A., Panizza, M., Electrochemical advanced oxidation processes: today and tomorrow. A review. *Environ. Sci. Pollut. Res.*, 21, 8336–8367, 2014, https://doi.org/10.1007/s11356-014-2783-1.
150. Brillas, E., Sirés, I., Oturan, M.A., Electro-Fenton Process and Related Electrochemical Technologies Based on Fenton's Reaction Chemistry. *Chem. Rev.*, 109, 6570–6631, 2009, https://doi.org/10.1021/cr900136g.
151. Ganiyu, S.O., Zhou, M., Martínez-Huitle, C.A., Heterogeneous electro-Fenton and photoelectro-Fenton processes: A critical review of fundamental principles and application for water/wastewater treatment. *Appl. Catal. B: Environ.*, 235, 103–129, 2018, https://doi.org/10.1016/j.apcatb.2018.04.044.

152. Brillas, E., A Review on the Degradation of Organic Pollutants in Waters by UV Photoelectro-Fenton and Solar Photoelectro-Fenton. *J. Braz. Chem. Soc.*, 25, 3, 393–417, 2014, https://doi.org/10.5935/0103-5053.20130257.
153. Nidheesh, P.V., Zhou, M., Oturan, M.A., An overview on the removal of synthetic dyes from water by electrochemical advanced oxidation processes. *Chemosphere*, 197, 210–227, 2018. https://doi.org/10.1016/j.chemosphere.2017.12.195.
154. Poza-Nogueiras, V., Rosales, E., Pazos, M., Sanromán, M.Á., Current advances and trends in electro-Fenton process using heterogeneous catalysts – A review. *Chemosphere*, 201, 399–416, 2018, https://doi.org/10.1016/j.chemosphere.2018.03.002.
155. Brillas, E. and Martínez-Huitle, C.A., Decontamination of wastewaters containing synthetic organic dyes by electrochemical methods. An updated review. *Appl. Catal. B: Environ.*, 166–167, 603–643, 2015, https://doi.org/10.1016/j.apcatb.2014.11.016.
156. Lianos, P., Review of recent trends in photoelectrocatalytic conversion of solar energy to electricity and hydrogen. *Appl. Catal. B: Environ.*, 210, 235–254, 2017, https://doi.org/10.1016/j.apcatb.2017.03.067.
157. Xu, P., Xu, H., Zheng, D., Simultaneous electricity generation and wastewater treatment in a photocatalytic fuel cell integrating electro-Fenton process. *J. Power Sources*, 421, 156–161, 2019, https://doi.org/10.1016/j.jpowsour.2019.03.033.
158. Lee, S.-L., Ho, L.-N., Ong, S.-A., Wong, Y.-S., Voon, C.-H., Khalik, W.F., Yusoff, N.A., Nordin, N., Enhanced electricity generation and degradation of the azo dye Reactive Green 19 in a photocatalytic fuel cell using ZnO/Zn as the photoanode. *J. Cleaner Prod.*, 127, 579–584, 2016, https://doi.org/10.1016/j.jclepro.2016.03.169.
159. Li, L., Xue, S., Chen, R., Liao, Q., Zhu, X., Wang, Z., He, X., Feng, H., Cheng, X., Performance characteristics of a membraneless solar responsive photocatalytic fuel cell with an air-breathing cathode under different fuels and electrolytes and air conditions. *Electrochim. Acta*, 182, 280–288, 2015, https://doi.org/10.1016/j.electacta.2015.09.090.
160. He, X., Chen, M., Chen, R., Zhu, X., Liao, Q., Ye, D., Zhang, B., Zhang, W., Yu, Y., A solar responsive photocatalytic fuel cell with the membrane electrode assembly design for simultaneous wastewater treatment and electricity generation. *J. Hazard. Mater.*, 358, 346–354, 2018, https://doi.org/10.1016/j.jhazmat.2018.07.031.
161. Bai, J., Wang, R., Li, Y., Tang, Y., Zeng, Q., Xia, L., Li, X., Li, J., Li, C., Zhou, B., A solar light driven dual photoelectrode photocatalytic fuel cell (PFC) for simultaneous wastewater treatment and electricity generation. *J. Hazard. Mater.*, 311, 51–62, 2016, https://doi.org/10.1016/j.jhazmat.2016.02.052.
162. Georgieva, J., Valova, E., Armyanov, S., Philippidis, N., Poulios, I., Sotiropoulos, S., Bi-component semiconductor oxide photoanodes for the photoelectrocatalytic oxidation of organic solutes and vapours: A short

review with emphasis to TiO2–WO3 photoanodes. *J. Hazard. Mater.*, 211–212, 30–46, 2012, https://doi.org/10.1016/j.jhazmat.2011.11.069.
163. Zeng, Q., Bai, J., Li, J., Li, L., Xia, L., Zhou, B., Sun, Y., Highly-stable and efficient photocatalytic fuel cell based on an epitaxial TiO2/WO3/W nanothorn photoanode and enhanced radical reactions for simultaneous electricity production and wastewater treatment. *Appl. Energy*, 220, 127–137, 2018, https://doi.org/10.1016/j.apenergy.2018.03.042.
164. Wang, D., Li, Y., Li Puma, G., Wang, C., Wang, P., Zhang, W., Wang, Q., Dye-sensitized photoelectrochemical cell on plasmonic Ag/AgCl @ chiral TiO 2 nanofibers for treatment of urban wastewater effluents, with simultaneous production of hydrogen and electricity. *Appl. Catal. B: Environ.*, 168–169, 25–32, 2015, https://doi.org/10.1016/j.apcatb.2014.11.012.
165. Paracchino, A., Laporte, V., Sivula, K., Grätzel, M., Thimsen, E., Highly active oxide photocathode for photoelectrochemical water reduction. *Nat. Mater.*, 10, 456–461, 2011, https://doi.org/10.1038/nmat3017.
166. Masudy-Panah, S., Siavash Moakhar, R., Chua, C.S., Tan, H.R., Wong, T.I., Chi, D., Dalapati, G.K., Nanocrystal Engineering of Sputter-Grown CuO Photocathode for Visible-Light-Driven Electrochemical Water Splitting. *ACS Appl. Mater. Interfaces*, 8, 1206–1213, 2016, https://doi.org/10.1021/acsami.5b09613.
167. Patel, M., Kim, H.-S., Patel, D.B., Kim, J., CuO photocathode-embedded semitransparent photoelectrochemical cell. *J. Mater. Res.*, 31, 3205–3213, 2016, https://doi.org/10.1557/jmr.2016.364.
168. Macías-Sánchez, J., Hinojosa-Reyes, L., Guzmán-Mar, J.L., Peralta-Hernández, J.M., Hernández-Ramírez, A., Performance of the photo-Fenton process in the degradation of a model azo dye mixture. *Photochem. Photobiol. Sci.*, 10, 332–337, 2011, https://doi.org/10.1039/c0pp00158a.
169. Lui, G., Jiang, G., Fowler, M., Yu, A., Chen, Z., A high performance wastewater-fed flow-photocatalytic fuel cell. *J. Power Sources*, 425, 69–75, 2019, https://doi.org/10.1016/j.jpowsour.2019.03.091.
170. Parmar, J., Jang, S., Soler, L., Kim, D.-P., Sánchez, S., Nano-photocatalysts in microfluidics, energy conversion and environmental applications. *Lab. Chip*, 15, 2352–2356, 2015, https://doi.org/10.1039/C5LC90047F.
171. Xia, M., Chen, R., Zhu, X., Liao, Q., An, L., Wang, Z., He, X., Jiao, L., A micro photocatalytic fuel cell with an air-breathing, membraneless and monolithic design. *Sci. Bull.*, 61, 1699–1710, 2016, https://doi.org/10.1007/s11434-016-1178-8.
172. Chow, C.-H. and Leung, K.S.-Y., Removing acesulfame with the peroxone process: Transformation products, pathways and toxicity. *Chemosphere*, 221, 647–655, 2019, https://doi.org/10.1016/j.chemosphere.2019.01.082.
173. Ding, Y., Wang, J., Xu, S., Lin, K.-Y.A., Tong, S., Oxygen vacancy of CeO2 improved efficiency of H2O2/O3 for the degradation of acetic acid in acidic solutions. *Sep. Purif. Technol.*, 207, 92–98, 2018, https://doi.org/10.1016/j.seppur.2018.06.027.

174. Jaafarzadeh, N., Barzegar, G., Ghanbari, F., Photo assisted electro-peroxone to degrade 2,4-D herbicide: The effects of supporting electrolytes and determining mechanism. *Process Saf. Environ. Prot.*, 111, 520–528, 2017, https://doi.org/10.1016/j.psep.2017.08.012.
175. Shen, W., Wang, Y., Zhan, J., Wang, B., Huang, J., Deng, S., Yu, G., Kinetics and operational parameters for 1,4-dioxane degradation by the photo-electro-peroxone process. *Chem. Eng. J.*, 310, 249–258, 2017, https://doi.org/10.1016/j.cej.2016.10.111.
176. Frangos, P., Wang, H., Shen, W., Yu, G., Deng, S., Huang, J., Wang, B., Wang, Y., A novel photoelectro-peroxone process for the degradation and mineralization of substituted benzenes in water. *Chem. Eng. J.*, 286, 239–248, 2016, https://doi.org/10.1016/j.cej.2015.10.096.
177. Joy, A.C., Gandhimathi, R., Niveditha, S.V., Ramesh, S.T., Nidheesh, P.V., Photoelectro-peroxone process for the degradation of reactive azo dye in aqueous solution. *Sep. Sci. Technol.*, 55, 2550–2559, 2020, https://doi.org/10.1080/01496395.2019.1634732.
178. Ahmadi, M. and Ghanbari, F., Degradation of organic pollutants by photoelectro-peroxone/ZVI process: Synergistic, kinetic and feasibility studies. *J. Environ. Manag.*, 228, 32–39, 2018, https://doi.org/10.1016/j.jenvman.2018.08.102.
179. Panizza, M., Delucchi, M., Sirés, I., Electrochemical process for the treatment of landfill leachate. *J. Appl. Electrochem.*, 40, 1721–1727, 2010, https://doi.org/10.1007/s10800-010-0109-7.
180. Ye, Z., Zhang, H., Zhang, X., Zhou, D., Treatment of landfill leachate using electrochemically assisted UV/chlorine process: Effect of operating conditions, molecular weight distribution and fluorescence EEM-PARAFAC analysis. *Chem. Eng. J.*, 286, 508–516, 2016, https://doi.org/10.1016/j.cej.2015.10.017.
181. Pellenz, L., Borba, F.H., Daroit, D.J., Lassen, M.F.M., Baroni, S., Zorzo, C.F., Guimarães, R.E., Espinoza-Quiñones, F.R., Seibert, D., Landfill leachate treatment by a boron-doped diamond-based photo-electro-Fenton system integrated with biological oxidation: A toxicity, genotoxicity and by products assessment. *J. Environ. Manag.*, 264, 110473, 2020, https://doi.org/10.1016/j.jenvman.2020.110473.
182. Renou, S., Givaudan, J.G., Poulain, S., Dirassouyan, F., Moulin, P., Landfill leachate treatment: Review and opportunity. *J. Hazard. Mater.*, 150, 468–493, 2008, https://doi.org/10.1016/j.jhazmat.2007.09.077.
183. Tauchert, E., Schneider, S., de Morais, J.L., Peralta-Zamora, P., Photochemically-assisted electrochemical degradation of landfill leachate. *Chemosphere*, 64, 1458–1463, 2006, https://doi.org/10.1016/j.chemosphere.2005.12.064.
184. Ghanbari, F., Wu, J., Khatebasreh, M., Ding, D., Lin, K.-Y.A., Efficient treatment for landfill leachate through sequential electrocoagulation, electrooxidation and PMS/UV/CuFe2O4 process. *Sep. Purif. Technol.*, 242, 116828, 2020, https://doi.org/10.1016/j.seppur.2020.116828.

185. Zhang, H., Li, Y., Wu, X., Statistical Experiment Design Approach for the Treatment of Landfill Leachate by Photoelectro-Fenton Process. *J. Environ. Eng.*, 138, 278–285, 2012, https://doi.org/10.1061/(ASCE)EE.1943-7870.0000448.
186. Klauck, C.R., Giacobbo, A., de Oliveira, E.D.L., da Silva, L.B., Rodrigues, M.A.S., Evaluation of acute toxicity, cytotoxicity and genotoxicity of landfill leachate treated by biological lagoon and advanced oxidation processes. *J. Environ. Chem. Eng.*, 5, 6188–6193, 2017, https://doi.org/10.1016/j.jece.2017.11.058.
187. Seibert, D., Henrique Borba, F., Bueno, F., Inticher, J.J., Módenes, A.N., Espinoza-Quiñones, F.R., Bergamasco, R., Two-stage integrated system photo-electro-Fenton and biological oxidation process assessment of sanitary landfill leachate treatment: An intermediate products study. *Chem. Eng. J.*, 372, 471–482, 2019, https://doi.org/10.1016/j.cej.2019.04.162.
188. Flores, N., Brillas, E., Centellas, F., Rodríguez, R.M., Cabot, P.L., Garrido, J.A., Sirés, I., Treatment of olive oil mill wastewater by single electrocoagulation with different electrodes and sequential electrocoagulation/electrochemical Fenton-based processes. *J. Hazard. Mater.*, 347, 58–66, 2018, https://doi.org/10.1016/j.jhazmat.2017.12.059.
189. Sharma, S. and Simsek, H., Treatment of canola-oil refinery effluent using electrochemical methods: A comparison between combined electrocoagulation + electrooxidation and electrochemical peroxidation methods. *Chemosphere*, 221, 630–639, 2019, https://doi.org/10.1016/j.chemosphere.2019.01.066.
190. Flores, N., Cabot, P.L., Centellas, F., Garrido, J.A., Rodríguez, R.M., Brillas, E., Sirés, I., 4-Hydroxyphenylacetic acid oxidation in sulfate and real olive oil mill wastewater by electrochemical advanced processes with a boron-doped diamond anode. *J. Hazard. Mater.*, 321, 566–575, 2017, https://doi.org/10.1016/j.jhazmat.2016.09.057.
191. Panizza, M., Fine Chemical Industry, Pulp and Paper Industry, Petrochemical Industry and Pharmaceutical Industry, in: *Electrochemical Water and Wastewater Treatment*, pp. 335–364, Elsevier: Amsterdam, Netherlands, 2018, https://doi.org/10.1016/B978-0-12-813160-2.00013-4.
192. Lalwani, J., Gupta, A., Thatikonda, S., Subrahmanyam, C., An industrial insight on treatment strategies of the pharmaceutical industry effluent with varying qualitative characteristics. *J. Environ. Chem. Eng.*, 8, 104190, 2020, https://doi.org/10.1016/j.jece.2020.104190.
193. Xu, D., Xiao, Y., Pan, H., Mei, Y., Toxic effects of tetracycline and its degradation products on freshwater green algae. *Ecotoxicol. Environ. Saf.*, 174, 43–47, 2019, https://doi.org/10.1016/j.ecoenv.2019.02.063.
194. Moreira, F.C., Soler, J., Alpendurada, M.F., Boaventura, R.A.R., Brillas, E., Vilar, V.J.P., Tertiary treatment of a municipal wastewater toward pharmaceuticals removal by chemical and electrochemical advanced oxidation processes. *Water. Res.*, 105, 251–263, 2016, https://doi.org/10.1016/j.watres.2016.08.036.

195. Behfar, R. and Davarnejad, R., Pharmaceutical wastewater treatment using UV-enhanced electro-Fenton process: Comparative study. *Water Environ. Res.*, 91, 1526–1536, 2019, https://doi.org/10.1002/wer.1153.
196. Fang, T., Liao, L., Xu, X., Peng, J., Jing, Y., Removal of COD and colour in real pharmaceutical wastewater by photoelectrocatalytic oxidation method. *Environ. Technol.*, 34, 779–786, 2013, https://doi.org/10.1080/09593330.2012.715760.
197. Martín de Vidales, M.J., Millán, M., Sáez, C., Cañizares, P., Rodrigo, M.A., Irradiated-assisted electrochemical processes for the removal of persistent pollutants from real wastewater. *Sep. Purif. Technol.*, 175, 428–434, 2017, https://doi.org/10.1016/j.seppur.2016.11.014.
198. GilPavas, E. and Correa-Sánchez, S., Optimization of the heterogeneous electro-Fenton process assisted by scrap zero-valent iron for treating textile wastewater: Assessment of toxicity and biodegradability. *J. Water Process Eng.*, 32, 100924, 2019, https://doi.org/10.1016/j.jwpe.2019.100924.
199. de Mello Florêncio, T., de Araújo, K.S., Antonelli, R., de Toledo Fornazari, A.L., da Cunha, P.C.R., da Silva Bontempo, L.H., de Jesus Motheo, A., Granato, A.C., Malpass, G.R.P., Photo-assisted electrochemical degradation of simulated textile effluent coupled with simultaneous chlorine photolysis. *Environ. Sci. Pollut. Res.*, 23, 19292–19301, 2016, https://doi.org/10.1007/s11356-016-6912-x.
200. Moraes, P.B., Pelegrino, R.R.L., Bertazzoli, R., Degradation of Acid Blue 40 dye solution and dye house wastewater from textile industry by photo-assisted electrochemical process. *J. Environ. Sci. Health Part A*, 42, 2131–2138, 2007, https://doi.org/10.1080/10934520701629591.
201. Salazar, R., Brillas, E., Sirés, I., Finding the best Fe2+/Cu2+ combination for the solar photoelectro-Fenton treatment of simulated wastewater containing the industrial textile dye Disperse Blue 3. *Appl. Catal. B: Environ.*, 115–116, 107–116, 2012, https://doi.org/10.1016/j.apcatb.2011.12.026.
202. Aravind, P., Subramanyan, V., Ferro, S., Gopalakrishnan, R., Eco-friendly and facile integrated biological-cum-photo assisted electrooxidation process for degradation of textile wastewater. *Water Res.*, 93, 230–241, 2016, https://doi.org/10.1016/j.watres.2016.02.041.
203. Manenti, D.R., Módenes, A.N., Soares, P.A., Espinoza-Quiñones, F.R., Boaventura, R.A.R., Bergamasco, R., Vilar, V.J.P., Assessment of a multistage system based on electrocoagulation, solar photo-Fenton and biological oxidation processes for real textile wastewater treatment. *Chem. Eng. J.*, 252, 120–130, 2014, https://doi.org/10.1016/j.cej.2014.04.096.
204. Bener, S., Atalay, S., Ersöz, G., The hybrid process with eco-friendly materials for the treatment of the real textile industry wastewater. *Ecol. Eng.*, 148, 105789, 2020, https://doi.org/10.1016/j.ecoleng.2020.105789.
205. Bharagava, R.N. and Mishra, S., Hexavalent chromium reduction potential of Cellulosimicrobium sp. isolated from common effluent treatment plant of

tannery industries. *Ecotoxicol. Environ. Saf.*, 147, 102–109, 2018, https://doi.org/10.1016/j.ecoenv.2017.08.040.

206. Lofrano, G., Meriç, S., Zengin, G.E., Orhon, D., Chemical and biological treatment technologies for leather tannery chemicals and wastewaters: A review. *Sci. Total Environ.*, 461–462, 265–281, 2013, https://doi.org/10.1016/j.scitotenv.2013.05.004.
207. Borba, F.H., Seibert, D., Pellenz, L., Espinoza-Quiñones, F.R., Borba, C.E., Módenes, A.N., Bergamasco, R., Desirability function applied to the optimization of the Photoperoxi-Electrocoagulation process conditions in the treatment of tannery industrial wastewater. *J. Water Process Eng.*, 23, 207–216, 2018, https://doi.org/10.1016/j.jwpe.2018.04.006.
208. Selvaraj, H., Aravind, P., George, H.S., Sundaram, M., Removal of sulfide and recycling of recovered product from tannery lime wastewater using photoassisted-electrochemical oxidation process. *J. Ind. Eng. Chem.*, 83, 164–172, 2020, https://doi.org/10.1016/j.jiec.2019.11.024.
209. Isarain-Chávez, E., de la Rosa, C., Godínez, L.A., Brillas, E., Peralta-Hernández, J.M., Comparative study of electrochemical water treatment processes for a tannery wastewater effluent. *J. Electroanal. Chem.*, 713, 62–69, 2014, https://doi.org/10.1016/j.jelechem.2013.11.016.
210. Moradi, M. and Moussavi, G., Enhanced treatment of tannery wastewater using the electrocoagulation process combined with UVC/VUV photoreactor: Parametric and mechanistic evaluation. *Chem. Eng. J.*, 358, 1038–1046, 2019, https://doi.org/10.1016/j.cej.2018.10.069.

9

Solar-Driven Photoelectrochemical Technologies for Wastewater Treatment and Green Hydrogen Production

Mohamed Mahmoud[1,2]

[1]*Water Pollution Research Department, National Research Centre, Dokki, Cairo, Egypt*
[2]*Faculty of Engineering, Galala University, Galala City, Suez, Egypt*

Abstract

Over the past few decades, there has been an urgent need to develop innovative water treatment technologies for eliminating pathogens and persistent micropollutants (e.g., pharmaceuticals and personal care products) that are commonly found in different waste streams with a wide range of concentrations (i.e., micrograms to nanograms per liter). More importantly, the scarce availability of valuable resources is the main reason for shifting our societal attention away from traditional waste management and disposal methods toward the recovery of resources from wastewater. A potential approach for water and wastewater treatment that generates net energy is solar-driven photoelectrochemical cells (PECs). Photoelectrochemical technologies open up opportunities to efficiently degrade recalcitrant wastewater and generate renewable energy compared to traditional advanced oxidation processes owing to their outstanding treatment efficiency and renewable energy production, mainly due to the synergistic impact of photochemical and electrolysis reactions in PEC. This book chapter provided a critical assessment of the current advances in photoelectrochemical technologies, including fundamental reaction mechanisms of different photoelectrochemical methods. The synergistic effects of various combined photochemical, electrochemical, and photoelectrochemical processes were also discussed. The performance capability of modified photoanode and cathode materials used in photoelectrochemical

Email: moh.mahmoud@nrc.sci.eg; mmali6384@gmail.com; m.mahmoud@gu.edu.eg
ORCID: https://orcid.org/0000-0002-8767-9743, Scopus ID: 57931010000

Sadanand Pandey, Elvis Fosso-Kankeu and Soumya Pandit (eds.) Synergy of Bio-Chemical Processes for Photocatalytic and Photoelectrochemical Wastewater Treatment, (245–264) © 2024 Scrivener Publishing LLC

reactors was described. Finally, the photoelectrochemical treatment of real wastewater for hydrogen production was reviewed.

***Keywords*:** Photoelectrochemical treatment, synergistic effects, recalcitrant wastewater, renewable energy, advanced oxidation processes, electrochemical treatment

9.1 Introduction

Conventional technologies for treating wastewater, which usually involve biological processes, have demonstrated high efficacy in degrading the majority of organic pollutants and suspended solids. However, these methods are inadequate in eliminating pathogens and persistent micropollutants (e.g., pharmaceuticals and personal care products) that are commonly found in different waste streams with a wide range of concentrations (ranging from micrograms to nanograms per liter) [1–3]. The application of these biological-based technologies necessitates the implementation of a post-treatment step in order to decrease the micropollutant concentration and the prevalence of pathogens to acceptable thresholds for diverse reuse purposes, including agricultural irrigation [4–6].

In recent decades, there has been significant interest in advanced oxidation processes (AOPs), which are treatment technologies designed to degrade and mineralize recalcitrant organic contaminants through multiple catalytic reactions taking advantage of the hydroxyl radical ($^{\bullet}OH$). Heterogeneous photocatalysis relies on using UV and visible light in conjunction with oxygen from the air to non-selectively decompose organic contaminants present in waste streams, representing an environmentally friendly technology for wastewater purification [7]. However, a bottleneck for technology scaling up is the fast recombination of charge carriers, which results in a significant decline in its efficacy [8].

In this context, photoelectrocatalysis, which is an example of an AOP, represents a promising technology for wastewater purification and renewable energy production. In a typical photoelectrochemical cell (PEC), a thin layer of photoactive semiconductor catalysts is deposited on the surface of a conductive electrode to act as either a photoanode or a photocathode (Figure 9.1) [9]. For example, the promotion of an electron from the valence band to the conduction band of a photoanode occurs only if the energy of the irradiation is equal to or greater than the energy of the photoanode's bandgap. This process leads to the creation of electron–hole pairs (e^-–h^+). A power supply is often used to provide the external bias required

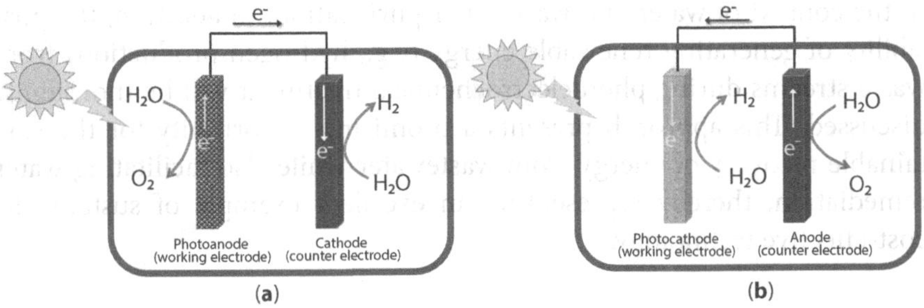

Figure 9.1 A schematic illustration of photoelectrochemical cells employing: (a) n-type semiconductor photoanodes and (b) p-type semiconductor photocathodes.

to facilitate this non-spontaneous reaction [10]. The ability to regulate the Fermi level of a semiconductor is instrumental in manipulating band bending, which, in turn, facilitates the effective segregation of electron-hole pairs and minimizes the possibility of their recombination. As holes migrate toward the surface of semiconductors, they facilitate the oxidation of water and generate hydroxyl radicals. Alternatively, they may directly oxidize organic compounds. Simultaneously, photogenerated electrons are conveyed to the counter electrode through an external circuit, where the reduction of an electron acceptor occurs [11]. In contrast to photocatalysis, photoelectrochemical processes result in enhanced process efficiency and higher production rates of reactive oxygen species (ROS) owing to the inhibition or slowing down of electron–hole pair recombination [12, 13]. When compared to conventional photocatalytic systems, photoelectrochemical processes involve the immobilization of the photocatalyst on a conductive substrate, thereby eliminating the need for separation steps to retrieve the photocatalyst [14].

Despite the promising use of photoelectrochemical processes for removing recalcitrant organic compounds and pathogens from contaminated water bodies, treating complex waste streams (such as real industrial wastewater) remains a challenging task, in which complex matrices can have adverse impacts on the photoelectrochemical catalytic degradation efficiency [15]. Hence, the majority of research studies focused on the detoxification of organic model compounds rather than real wastewater [16]. More importantly, the photoelectrochemical treatment becomes more complicated when attempting to target both organic matter and pathogens concurrently, as the efficiency of the process may be reduced due to competition for the generated ROS. This book chapter will examine crucial elements pertaining to the design and expansion of PECs utilized in the context of water and wastewater purification. In addition, the possibility of generating renewable energy (e.g., hydrogen production) from waste streams during photoelectrochemical treatment will be thoroughly discussed. This approach presents a promising opportunity for the sustainable recovery of energy from wastewater while also facilitating water remediation, thereby representing an excellent example of sustainable, cost-effective technology.

9.2 Photoelectrochemical Catalysts

In a typical photocatalytic system, semi-conductive nanomaterials play a crucial role in wastewater detoxification because their band edge potentials enable the efficient generation of hydroxyl and superoxide ion radicals. However, for photoelectrochemical processes, the significance of generating both radicals is somewhat diminished as the auxiliary electrode can effectively facilitate a redox reaction. The production of ROS through photoelectrochemical reactions could be enhanced by the application of an external bias, thereby enabling the completion of the redox cycle (Equations 9.1 to 9.4) [13]. Materials possessing a lower bandgap energy (E_{bg}) may be utilized in this context, whereas the counter electrode can facilitate the transfer of electrons to the solution.

$$OH^- + h^+ \rightarrow {}^{\bullet}OH \ (E = +2.29 \text{ V vs. SHE}) \tag{9.1}$$

$$O_2 + 2e^- + 2H^+ \rightarrow H_2O_2 \ (E = +0.281 \text{ vs. SHE}) \tag{9.2}$$

$$O_2 + e^- + H^+ \rightarrow \bullet HO_2 \ (E = -0.05 \text{ vs. SHE}) \tag{9.3}$$

$$O_2 + e^- \rightarrow O_2^{\bullet -} \ (E = -0.33 \text{ vs. SHE}) \tag{9.4}$$

Most studies on photoelectrochemical systems employ n-type materials as photoanodes, primarily because p-type photocathode materials are known to be unstable [17]. Considerable research efforts have been directed toward improving the performance of photoanodes. This has been achieved through the utilization of materials with lower bandgaps, heterostructures, and/or co-catalysts. These enhancements aim to improve the photoanodes' ability to absorb visible light, enhance stability, facilitate charge transfer, and reduce the oxidation reaction overpotential. Enhancing the overall performance is contingent upon the improvement of photoelectrodes.

The utilization of lower bandgap materials and/or heterostructures has been extensively explored for enhancing solar performance by enabling the absorption of more visible light. However, the majority of research in this area has been directed toward solar energy harvesting rather than water treatment [18]. Furthermore, insufficient attention has been devoted to the function of the auxiliary electrodes in photoelectrochemical systems because they have the potential for producing supplementary ROS and influencing the efficacy of PEC.

Numerous research studies have focused on the advancement of semiconductor materials that can be utilized in photoelectrochemical-based systems for diverse purposes, including the decomposition of pollutants, the deactivation of microorganisms, and the production of hydrogen [19–22]. In order to meet the requirements of photoelectrochemical applications, semiconductor materials are expected to possess appropriate bandgap energies (E_{bg}), high mobility and diffusion of charge carriers, efficient light harvesting in both UV and visible regions, outstanding stability and activity, and relatively low price [15, 22]. In this context, there are a variety of semiconductor-based photocatalysts with different electronic structures and properties that have been widely used for photoelectrochemical applications [23, 24]. Among different promising photocatalysts, TiO_2, ZnO, and WO_3 are frequently utilized in photoelectrochemical applications due to their outstanding photocatalytic characteristics compared to other photocatalysts, enhanced chemical stability, relatively low cost, reusability, being environmentally friendly, and appropriate bandgap [12, 22].

9.3 Photoelectrochemical Cell Design and Operation for Wastewater Treatment

Photoelectrochemical processes often degrade organic compounds and inactivate pathogens in contaminated water through either direct oxidation/reduction, indirect degradation through the generation of ROS, or a combination of both [25]. Several reactor configurations have been documented in the literature. However, a common feature among these different reactor configurations underscores the necessity for having a high electrode surface area to reactor volume ratio with minimal electrode distance. The degradation rates can be enhanced by improving pollutants' mass transport as well as supplying enough quantity of oxygen for enhancing the ROS generation. Ideally, the commercialization and scaling-up of photoelectrochemical technology require that PEC reactors are operated in a continuous-flow mode, wherein wastewater is efficiently treated to meet the required limits for water reuse, especially for agricultural purposes [26, 27]. Furthermore, the incorporation of aeration within PEC reactors serves a two-fold function: ensuring a high concentration of oxygen, which is required for the reductive pathway to generate ROS, and enhancing mass transport. In a recent study, Kim *et al.* [28] investigated the degradation of benzophenone-3 and *Candida parapsilosis* in a PEC reactor in the presence of O_2 and O_3. They revealed that the utilization of O_3 led to reduced figures of merit in comparison to O_2. It is worth mentioning that their analysis solely takes into account the electricity consumption per unit of reduction, without factoring in the initial investment cost.

Given that the PEC reactors necessitate an irradiation source, there are two main alternatives: solar or artificial light sources. Thus, the utilization of energy-efficient, relatively cheap light-emitting diodes (LEDs) allows for the operation of PEC reactors at any given time. However, it is important to note that there are supplementary expenses, in terms of initial investment and ongoing operation costs [29]. For example, Montenegro-Ayo *et al.* [30] presented a novel PEC reactor design for point-of-use water reuse. In their experiments, they employed a TiO_2 nanotube-based photoanode and a UVA/LED (i.e., 365 nm) to inactivate *E. coli* in contaminated water streams. Within only 6 min of operation, they observed almost complete bacterial inactivation, implying promising applicability of UV-LED–based PEC system for bacterial disinfection.

On the other hand, the use of real irradiation light represents a much cheaper option for achieving high efficiency of waste and wastewater treatment compared to simulated irradiation sources. Despite the potential benefits, the implementation of this approach may pose certain challenges, including the need for suitable reactor configurations and the variability of solar radiation levels, which can impact the duration of the treatment process. McMichael *et al.* [20] conducted a study on reactor design to test the efficacy of inactivating *E. coli* and *P. aeruginosa* under real solar irradiation. The study revealed a significant finding that the photocurrent of a titanium nanotube photoanode exhibited a linear correlation with the variation in UV intensity. This finding holds potential for the development of a quality assurance technique for the use of real solar irradiation.

Although the main goal of PEC systems is to detoxify organic pollutants and pathogens in water and wastewater, the majority of studies only report results based on laboratory-scale trials [31]. The dearth of scaled-up PEC can potentially be attributed to the inadequacy of predictive accurate models, which are capable of facilitating the design and development of PEC systems [32]. A limited number of manuscripts have been published wherein an empirical model is presented for the elimination of organic pollutants [33]. Several models have been documented for the purpose of designing and optimizing photoelectrochemical reactors that facilitate water splitting to generate hydrogen and oxygen [34]. These models are commonly constructed utilizing advanced multi-physics software due to their capability to couple and solve the various physical phenomena that dictate the process, including, but not limited to, radiation absorption, electrochemistry, and fluid flow. The adaptation of current models for wastewater treatment can be achieved through the development of intrinsic kinetic models [35], which involve the intrinsic kinetic parameters estimation as well as chemical reaction rates [36]. In another study, Salmerón *et al.* [37] proposed a water-splitting model that has the potential to be modified for the purpose of simulating the evaluation of various electrode geometries and configurations, as well as their resulting performance. This can be achieved by evaluating the spatial patterns and arrangements of potential, photon flux, and electric current density.

9.4 Application of Photoelectrochemical Systems for Real Wastewater Treatment

Over the past few years, a considerable number of research studies have employed impractical aqueous solutions, such as NaCl and Na_2SO_4. Further research is necessary to expand the scale of PEC systems, utilizing real water and wastewater samples. Real water matrices are anticipated to comprise a blend of organic and inorganic compounds that will contend for ROS generation, thereby leading to reduced reaction rates [31]. The inclusion of chlorine is advantageous as it yields reactive chlorine species that can effectively inactivate pathogens and detoxify with organic pollutants [35, 36]. Typically, the majority of research studies report using only single contaminant degradation; however, there is a desire to investigate the impact of co-occurring pollutants, such as organic micropollutants and pathogens. For instance, Salmerón *et al.* [37] evaluated the efficiency of a PEC reactor to concurrently remove *E. coli* and three organic pollutants (i.e., terbutryn, clorfenvinphos, and diclofenac) in real water samples. They observed a reasonably high efficiency for achieving bacterial inactivation and pollutant degradation without the need to alter the supporting electrolytes. Although, further investigation is necessary to bring PEC systems from laboratory-scale systems into real application as pilot systems operating on real wastewater.

Landfill leachate is produced as a result of the physical, chemical, and biological breakdown of municipal solid waste in engineered landfills, as well as the infiltration and drainage of water through the waste materials [38–40]. Landfill leachate typically contains high levels of recalcitrant organic pollutants, inorganic salts, heavy metals, and ammonia nitrogen, with a biochemical oxygen demand-to-chemical oxygen demand ratio of as low as 0.004, necessitating the implementation of AOPs in order to attain satisfactory detoxification efficiency [38, 39]. Given that landfill leachate has a dark color, the photo-based AOPs approach, including photoelectrochemical treatment, seems to be inefficient at reducing the acute toxicity and cytotoxicity of landfill leachate [41]. In a recent study, Ye *et al.* [42] conducted an assessment of the applicability of photoelectrochemical treatment of biologically pre-treated landfill leachate. At a slightly acidic pH (i.e., 5.0) and a relatively high current density of 60 mA cm^{-2}, a removal efficiency of approximately 87% for NH_3–N and 77% for COD was achieved without the use of a supporting electrolyte. During an 8-h treatment, the total energy consumption was ~216.5 kW h/kg COD with a current efficiency of 17.5%. Throughout the course of leachate treatment, a significant portion of NH_3-N converted into N_2, whereas a minor quantity of NO_2^-, NO_3^-, and chloramines persisted within the electrolyte. In another study, Zhao and colleagues [43] used a pilot, continuous-flow PEC reactor to remove organic matter present in biologically pre-treated landfill leachate. Within a 2.5-h treatment duration, they observed 41.6% removal of total organic carbon, 74.1% removal of COD, and 94.5% removal of ammonium at an applied bias of 67.1 mA cm^{-2}. More importantly, they revealed that the efficacy of organic contaminant mineralization into small molecular acids was impacted by UV irradiation.

Pharmaceutical wastewater contains a diverse range of compounds, including synthetic and natural organic substances, inorganic materials, catalysts, and solvents, with a relatively low biodegradability ratio [44]. More seriously, the presence of xenobiotics, solids, and active antibiotic residues in pharmaceutical wastewater results in a high level of toxicity [45–47]. Several conventional treatment techniques may not be appropriate in mitigating the adverse effects on both human and animal populations. This is evidenced by the emergence of antibiotic-resistant bacteria and genes [48]. Therefore, it is imperative to employ more advanced treatment methods to manage pharmaceutical wastewater. For instance, Fang *et al.* [49] used a PEC reactor equipped with a 250-W high-pressure mercury lamp as a source of irradiation, TiO_2-coated on Ni photoanode, and multi-walled carbon nanotubes air cathode. After a 2-h treatment, the untreated wastewater with a COD concentration of 3m150 mg L^{-1} experienced a 93.5% reduction in COD and a 78.5% reduction in color. The significant contributor to COD removal was identified as the depletion of Cl^-. Consequently, the increase in NaCl concentration as well as applied electrical bias led to an enhanced treatment efficiency, in terms of COD removal. However, the implementation of a single technique may at times prove insufficient in effectively eliminating recalcitrant organic compounds from aqueous solutions. In this context, de Vidales *et al.* [50] combined UV photolysis and conductive diamond electrochemical oxidation in one reactor to achieve high efficiency of pharmaceutical wastewater. In another study, Dindaş *et al.* [51] aimed to assess the efficacy of different combinations of electrocoagulation (EC), electro-Fenton (EF), and photocatalysis in the treatment of pharmaceutical wastewater. The experimental results indicated that the combination of EF (at a current density of 5 mA cm^{-2} for 1 h) and photocatalysis was the most effective in eliminating ~ 64% of total organic carbon (TOC), ~ 70% of chemical oxygen demand (COD), and ~ 98% of biological oxygen demand (BOD).

Textile wastewater exhibits highly variable complex composition due to the presence of diverse types of dyes, salts, pesticides, surfactants, and coating agents. Therefore, traditional treatment technologies are not viable to ensure the elimination of the majority of organic and inorganic pollutants. GilPavas *et al.* [52] developed a PEC reactor equipped with a monopolar arrangement of BDD/Ti electrodes for treating textile wastewater. Within 15 min, they observed an 83% reduction in COD while simultaneously increasing the biodegradability ratio (i.e., BOD_5/COD) to a level exceeding 0.4.

The tanning industry, being a completely aqueous process, utilizes a significant quantity of water and discharges approximately 30 L to 35 L of contaminated effluent for each kg of skin/hide processed [53–55]. The discharge of substances that result from the conversion of unprocessed or partially processed hides into marketable products, which consist of acids/bases, chromium salts, solvents, tannins, dyes, and sulfides, has been commonly documented [56]. Consequently, tannery wastewater is characterized by a diverse composition with high levels of organic and inorganic compounds (e.g., COD = 1 g L^{-1} to 58 g L^{-1}, total dissolved solids = 22 g L^{-1} to 67 g L^{-1}, Cr^{6+} = 0.01 mg L^{-1} to 4.24 mg L^{-1}, and sulfides = 0.025 g L^{-1} to 3.3 g L^{-1} [47, 49]. Selvaraj *et al.* [57] ascertained the significance of photoelectrochemical treatment of tannery effluent through the utilization of an electrochemical system consisting of a Ti/IrO_2–RuO_2–TiO_2 photoanode and a titanium cathode. Upon exposure to UV light illumination with a maximum wavelength of 254 nm, complete elimination of sulfide and 92% reduction in COD were observed, along with a 70% decrease in TOC. Conversely, in the absence of UV irradiation, the treatment was found to be ineffective. In another study conducted by Isarain-Chávez *et al.* [58], they examined the efficacy of a BDD-based photoelectro-Fenton (PEF) process in treating tannery wastewater. Throughout the procedure, the oxidation capacity of Boron-doped diamond (BDD), hydroxyl radicals (•OH), and ultraviolet A radiation collectively facilitated the efficient mineralization of organic contaminants present in the tannery effluent.

9.5 Hydrogen Production During Photoelectrochemical Treatment of Wastewater

Despite the vast reserves of fossil fuels that could potentially meet the energy needs of our society for several centuries, adhering to the current energy supply approach would result in an unprecedented elevation in the concentration of carbon dioxide (CO_2) during the coming few decades. Therefore, there is an escalating need for developing renewable and sustainable energy sources on a global scale [59–63]. Hydrogen has been extensively researched as a promising clean energy carrier for the future, owing to its greater specific energy density, superior efficiency, and environmentally friendly combustion process [64–66]. In addition to being able to effectively eliminate organic pollutants and pathogens, the photoelectrochemical water-splitting process has proven to be effective in the production of H_2 and is a technology with great potential for the production of sustainable energy while also yielding oxygen as a secondary product [67, 68]. In a typical PEC reactor, photoanodes are responsible for instigating the degradation of pollutants through enhanced photocatalytic activity, whereas cathodes have a more negative potential than the standard hydrogen electrode potential of 0 V, which is utilized for hydrogen production. Furthermore, sustainable photoelectrochemical processes are dependent on cost-effective external energy sources such as solar irradiation, wind energy, and biomass.

Furthermore, the photoanode properties have a significant impact on hydrogen production and organic pollutant degradation. Moreover, photoanode efficiency is contingent upon the bandgap of the material, whereby materials with smaller bandgaps are readily activated, thereby facilitating expeditious charge transport. The utilization of organic compounds and wastewater as sacrificial agents in aqueous media has the potential to enhance the photocatalytic efficiency of catalysts by impeding charge recombination. The performance of PECs was enhanced by the presence of organic waste in water samples [69]. This improvement was attributed to the organic waste acting as a sacrificial agent, resulting in increasing the generated photocurrent. H_2 generation was solely detected when exposed to visible light illumination and an external bias of 0.8 V and 1.6 V, resulting in an average H_2 generation of 0.36 mmol h^{-1}. The results of the experiment indicate that ethanol is a more efficient fuel source for hydrogen production when compared to glycerol and sorbitol. This can be attributed to the fact that ethanol is more extensive with the possibility to produce a greater number of H^+ per carbon atom. On the other hand, the nature of cathode materials is a crucial factor in determining their efficiency for reducing protons to facilitate spontaneous H_2 evolution. The choice of cathodic materials is influenced by various factors, such as cost, environmental sustainability, chemical and thermal stability, and H_2 production efficiency [70]. Among several cathodic materials, platinum- and copper-based catalysts are the most commonly used catalysts for several electrochemical energy storage and production technologies, including PEC [71–76]. Furthermore, there is ongoing research to alter the compositional structure of the cathode materials in order to enhance the efficacy of the PEC systems through the amelioration of charge separation and mitigation of photocorrosion.

Chen *et al.* [77] reported the observation of a synergistic effect in a PEC employing C-N co-doped TiO_2 nanotube arrays. The process involves the simultaneous generation of H_2 and removal of pollutants through electrochemical oxidation under an externally applied electric current. The process of photoelectrochemical catalysis is facilitated by the application of electric current, which serves to augment charge separation and promote electron transfer to the cathode. In another study, Zhao *et al.* [78] investigated the utilization of a photoelectrochemical catalytic approach to achieve concurrent wastewater treatment and H_2 production. This was accomplished through the implementation of a hybrid porous g-C_3N_4 anchored on reduced graphene oxide, which was subsequently loaded onto Ni foams as a photoanode. According to their report, organic contaminates functioned as a sacrificial agent, whereby the photogenerated holes underwent a reaction with them, resulting in the production of H_2 at the platinum-based cathode through the remaining electrons. Upon exposure to visible light, a ~21.6% reduction in the Rhodamine dye concentration was observed, which coincided with the maximum hydrogen production rate of 5.8 μmol h^{-1} cm^{-2}. Moreover, the electrode's efficiency exhibited remarkable stability over the course of three experimental cycles with consistent hydrogen evolution.

More importantly, several structural alterations, such as the introduction of metal doping, sensitization with semiconductors of lower bandgap, and catalyst modification, have been devised to enhance the characteristics of photoanode materials [79, 80]. Wang *et al.* [81] employed a dye-sensitized PEC for the purpose of H_2 production. In this study, they employed a novel photoanode composed of nanostructured plasmonic Ag/AgCl@chiral TiO_2 nanofibers, whereas the electrolyte used in the experiment consisted of urban wastewater containing an estrogen (17-β-ethynylestradiol) and a heavy metal (Cu^{2+}). They noticed a complete degradation of pollutants, which was accompanied by a high solar energy conversion rate. The improved efficiency was ascribed to the synergistic interplay between the chiral nanofibers of TiO_2 and Ag nanoparticles present in the photoanode. This interplay effectively impeded the recombination process, thereby facilitating the rapid degradation of pollutants and consequent generation of H_2 at the cathode. Koo *et al.* [82] conducted a study wherein they were able to achieve the degradation of pollutants and the production of H_2 simultaneously. This was made possible through the utilization of electrochromic titania nanotube arrays, specifically the Blue TiO_2 nanotubes (TNTs) as a photoanode. The application of cathodic polarization on TNTs resulted in a notable increase in electrical conductivity. The resulting photocurrent was then utilized by the photocathode to facilitate the production of hydrogen gas. The electrochemical process utilized for the surface modification of TNTs resulted in an enhancement of the photoanode's overall properties. This modification led to an increase in visible light absorption and a doubling of H_2 generation when compared to the performance of unmodified TNTs. The light source is a crucial aspect that must be taken into account for the PEC process. The utilization of high-capacity and sustainable energy sources is favored due to their ability to readily initiate the generation of electron–hole pairs at the photoanode. The utilization of sunlight as a feasible irradiation source in the PEC process can effectively address the constraints associated with photocatalyst materials that require uninterrupted light energy input [83].

9.6 Conclusions

In order to enhance the efficacy of PEC processes, it is crucial to undertake an appropriate selection of electrode materials, optimize the photoelectrode (e.g., TiO_2, ZnO, SnO_2, WO_3, and Fe_2O_3), and design the reactor accordingly. However, the counter electrode, typically composed of stainless steel, carbon paper/felt, platinum, or titanium, is not commonly analyzed. In order to successfully transition photoelectrochemical technologies from laboratory to commercial application, it is imperative to take into account the diverse range of accomplishments that have been realized through the utilization of various photoelectrochemical reactors. For instance, the scaling up and commercialization of the PEC technology will require the active surface area of the photoelectrode to the overall reactor volume ratio to be maximized, reduce the separation distance between electrodes, and minimize mass transfer. The majority of published articles document laboratory experiments conducted under highly controlled conditions with respect to water matrix and pollution. Consequently, it is necessary to conduct research utilizing both synthetic and actual wastewater to evaluate the feasibility of these systems in real-world scenarios. The inclusion of such analysis will provide perspectives on potential undesired reactions and the interplay among various contaminants and inorganic substances within the PEC. The utilization of PEC reactors in the context of wastewater remediation and hydrogen production presents a promising dual solution for both water remediation and energy production. This approach offers a cost-effective technology, which includes reduced capital costs, effective cost per unit treated, and potential for real-world adoption when compared to existing technologies. However, the effective application and exploitation of PEC technology will require close collaboration of electrochemists, engineers, and analytical chemists. In order to achieve this objective, it is imperative to use real waste streams rather than simple substrates, develop and construct stable, cost-effective, and durable photoelectrodes on an industrial scale, and conduct techno-economic analyses to establish their feasibility in comparison to existing wastewater treatment technologies.

Acknowledgments

The author wishes to acknowledge the the Science, Technology & Innovation Funding Authority, Egypt for funding this study (grant number 33445).

References

1. Pistocchi, A., Andersen, H.R., Bertanza, G., Brander, A., Choubert, J.M., Cimbritz, M., Drewes, J.E., Koehler, C., Krampe, J., Launay, M., Nielsen, P.H., Obermaier, N., Stanev, S., Thornberg, D., Treatment of micropollutants in wastewater: Balancing effectiveness, costs and implications. *Sci. Total Environ.*, 850, 157593, 2022.

2. Rafieenia, R., Sulonen, M., Mahmoud, M., El-Gohary, F., Rossa, C.A., Integration of microbial electrochemical systems and photocatalysis for sustainable treatment of organic recalcitrant wastewaters: Main mechanisms, recent advances, and present prospects. *Sci. Total Environ.*, 824, 153923, 2022.
3. Mahmoud, M. and El-Liethy, M.A., Three-dimensional, flow-through silver-magnetite nanocomposite–modified reactive electrochemical system with slow silver ions release for efficient bacterial disinfection of sewage. *J. Environ. Chem. Eng.*, 10, 106985, 2022.
4. Deng, Y. and Zhao, R., Advanced oxidation processes in wastewater treatment. *Curr. Pollut. Rep.*, 1, 167–176, 2015.
5. Youssef, Y.A., Abuarab, M.E., Mahrous, A., Mahmoud, M., Enhanced degradation of ibuprofen in an integrated constructed wetland-microbial fuel cell: treatment efficiency, electrochemical characterization, and microbial community dynamics. *RSC Adv.*, 13, 29809–29818, 2023.
6. El-Kamah, H. and Mahmoud, M., Performance evaluation of sequencing batch reactor for beverage industrial wastewater treatment. *Water Environ. Res.*, 84, 155–161, 2012, 8.
7. Rafieenia, R., Mahmoud, M., El-Gohary, F., Rossa, C.A., The degradation of glyphosate is enhanced in a microbial fuel cell: Electrochemical performance, degradation efficiency, and analysis of the anodic microbial community. *Sustain. Energy Technol. Assess.*, 54, 102805, 2022.
8. Lianos, P., Production of electricity and hydrogen by photocatalytic degradation of organic wastes in a photoelectrochemical cell: The concept of the Photofuelcell: A review of a re-emerging research field. *J. Hazard. Mater.*, 185, 575–590, 2011.
9. Pan, Y., Zhang, H., Zhang, B., Gong, F., Feng, J., Huang, H., Vanka, S., Fan, R., Cao, Q., Shen, M., Li, Z., Zou, Z., Xiao, R., and Chu, S., Renewable formate from sunlight, biomass and carbon dioxide in a photoelectrochemical cell. *Nat. Commun.*, 14, 1013, 2023.
10. Li, J. and Wu, N., Semiconductor-based photocatalysts and photoelectrochemical cells for solar fuel generation: a review. *Catal. Sci. Technol.*, 5, 1360–1384, 2015.
11. Coha, M., Farinelli, G., Tiraferri, A., Minella, M., Vione, D., Advanced oxidation processes in the removal of organic substances from produced water: potential, configurations, and research needs. *Chem. Eng. J.*, 414, 128668, 2021.
12. Garcia-Segura, S. and Brillas, E., Applied photoelectrocatalysis on the degradation of organic pollutants in wastewaters. *J. Photochem. Photobiol. C: Photochem. Rev.*, 12, 1-35, 2017.
13. Li, X., Yu, J., Jaroniec, M., Hierarchical photocatalysts. *Chem. Soc. Rev.*, 45, 2603–2636, 2016.
14. Kusmierek, E., Semiconductor electrode materials applied in photoelectrocatalytic wastewater treatment—an overview. *Catalysts*, 10, 439, 2020.

15. Brillas, E. and Martinez-Huitle, C.A., Decontamination of wastewaters containing synthetic organic dyes by electrochemical methods. An updated review. *Appl. Catal. B: Environ.*, 166/167, 603–643, 2015.
16. Amaterz, E., Tara, A., Bouddouch, A., Taoufyq, A., Bakiz, B., Benlhachemi, A., Jbara, O., Photo-electrochemical degradation of wastewaters containing organics catalysed by phosphate-based materials: a review. *Rev. Environ. Sci. Bio/Technol.*, 9, 843–872, 2020.
17. Huang, Q., Ye, Z., Xiao, X., Recent progress in photocathodes for hydrogen evolution. *J. Mater. Chem. A*, 3, 15824–15837, 2015.
18. Hendi, A.H., Osmanm, A.M., Khan, I., Saleh, T.A., Kandiel, T.A., Qahtan, T.F., Hossain, M.K., Visible light-driven photoelectrocatalytic water splitting using z scheme Ag-decorated MoS_2/RGO/$NiWO_4$ heterostructure. *ACS Omega*, 5, 31644–31656, 2020.
19. McMichael, S., Fernández-Ibáñez, P., Byrne, J.A., A review of photoelectrocatalytic reactors for water and wastewater treatment. *Water*, 13, 1198, 2021.
20. McMichael, S., Waso, M., Reyneke, B., Khan, W., Byrne, J.A., Fernandez-Ibanez, P., Electrochemically assisted photocatalysis for the disinfection of rainwater under solar irradiation. *Appl. Catal. B*, 281, 119485, 2021.
21. Collivignarelli, M.C., Abbà, A., Miino, M.C., Caccamo, F.M., Torretta, V., Rada, E.C., Sorlini, S., Disinfection of wastewater by UV-based treatment for reuse in a circular economy perspective. Where are we at? *Int. J. Environ. Res. Public Health*, 18, 1–24, 2021.
22. Wang, Y., Zu, M., Zhou, X., Lin, H., Peng, F., Zhang, S., Designing efficient TiO_2-based photoelectrocatalysis systems for chemical engineering and sensing. *Chem. Eng. J.*, 381, 122605, 2020.
23. Bhunia, P. and Dutta, K., Photocatalysts and photoelectrocatalysts in fuel cells and photofuel cells, in: *Green Photocatalysts for Energy and Environmental Process. Environmental Chemistry for a Sustainable World*, vol. 36, S. Rajendran, M. Naushad, L. Ponce, E. Lichtfouse (Eds.), Springer, Cham, 2020.
24. Arunachalam, P. and Al Mayouf, A.M., Photoelectrochemical water splitting, in: *Micro and Nano Technologies, Noble Metal-metal Oxide Hybrid Nanoparticles*, S. Mohapatra, T.A. Nguyen, P. Nguyen-Tri (Eds.), pp. 585–606, Woodhead Publishing, 2019. https://doi.org/10.1016/B978-0-12-814134-2.00028-0
25. Feng, Z., Tian, Q., Yang, Q., Zhou, Y., Zhao, H., Selectively photoelectrocatalytic reduction of oxygen to hydroxyl radical and singlet oxygen: mechanism and validation in coal wastewater. *Appl. Catal. B*, 286, 119908, 2021. https://doi.org/10.1016/j.apcatb.2021.119908
26. El-Qelish, M. and Mahmoud, M., Overcoming organic matter limitation enables high nutrient recovery from sewage sludge reject water in a self-powered microbial nutrient recovery cell. *Sci. Total Environ.*, 802, 149851, 2022.

27. Wahaab, R.A., Mahmoud, M., van Lier, J.B., Toward achieving sustainable management of municipal wastewater sludge in Egypt: The current status and future prospective. *Renew. Sustain. Energy Rev.*, 127, 109880, 2020.
28. Kim, J.Y.U., Bessegato, G.G., de Souza, B.C., da Silva, J.J., Zanoni, M.V.B., Efficient treatment of swimming pool water by photoelectrocatalytic ozonation: inactivation of Candida parapsilosis and mineralization of benzophenone-3 and urea. *Chem. Eng. J.*, 378, 122094, 2019.
29. Kangkun, N. and Ponchio, C., Photoelectrodeposition of $BiVO_4$ layer on FTO/WO_3 photoanodes for highly efficient photoelectrocatalytic chemical oxygen demand sensor applications. *Appl. Surf. Sci.*, 526, 146686, 2020.
30. Montenegro-Ayo, R., Barrios, A.C., Mondal, I., Bhagat, K., Morales-Gomero, J.C., Abbaszadegan, M., Westerhoff, P., Perreault, F., Garcia-Segura, S., Portable point-of-use photoelectrocatalytic device provides rapid water disinfection. *Sci. Total Environ.*, 737, 140044, 2020.
31. Rajput, H., Changotra, R., Sangal, V.K., Dhir, A., Photoelectrocatalytic treatment of recalcitrant compounds and bleach stage pulp and paper mill effluent using Au-TiO_2 nanotube electrode. *Chem. Eng. J.*, 408, 127287, 2021.
32. Bedoya-Lora, F.E., Hankin, A., Kelsall, G.H., En route to a unified model for photo-electrochemical reactor optimisation. I-photocurrent and H_2 yield predictions. *J. Mater. Chem. A*, 5, 22683–22696, 2017.
33. Turolla, A., Bestetti, M., Antonelli, M., Optimization of heterogeneous photoelectrocatalysis on nanotubular TiO_2 electrodes: reactor configuration and kinetic modelling. *Chem. Eng. Sci.*, 182, 171–179, 2018.
34. Hankin, A., Bedoya-Lora, F.E., Ong, C.K., Alexander, J.C., Petter, F., Kelsall, G.H., From millimetres to metres: the critical role of current density distributions in photo-electrochemical reactor design. *Energy Environ. Sci.*, 10, 346–360, 2017.
35. Farhat, A., Keller, J., Tait, S., Radjenovic, J., Assessment of the impact of chloride on the formation of chlorinated by-products in the presence and absence of electrochemically activated sulfate. *Chem. Eng. J.*, 330, 1265–1271, 2017.
36. Casado, C., Marugán, J., Timmers, R., Muñoz, M., van Grieken, R., Comprehensive multiphysics modeling of photocatalytic processes by computational fluid dynamics based on intrinsic kinetic parameters determined in a differential photoreactor. *Chem. Eng. J.*, 310, 368–380, 2017.
37. Salmerón, I., Sharma, P.K., Polo-López, M.I., Tolosana, A., McMichael, S., Oller, I., Byrne, J.A., Fernández-Ibáñez, P., Electrochemically assisted photocatalysis for the simultaneous degradation of organic micro-contaminants and inactivation of microorganisms in water. *Process Saf. Environ. Prot.*, 147, 488–496, 2021.
38. Panizza, M., Delucchi, M., Sirés, I., Electrochemical process for the treatment of landfill leachate. *J. Appl. Electrochem.*, 40, 1721–1727, 2010.

39. Mahmoud, M., Parameswaran, P., Torres, C.I., Rittmann, B.E., Relieving the fermentation inhibition enables high electron recovery from landfill leachate in a microbial electrolysis cell. *RSC Adv.*, 6, 6658–6664, 2016.
40. Mahmoud, M., Parameswaran, P., Torres, C.I., Rittmann, B.E., Fermentation pre-treatment of landfill leachate for enhanced electron recovery in a microbial electrolysis cell. *Bioresour. Technol.*, 151, 151–158, 2014.
41. Klauck, C.R., Giacobbo, A., de Oliveira, E.D.L., da Silva, L.B., Rodrigues, M.A.S., Evaluation of acute toxicity, cytotoxicity and genotoxicity of landfill leachate treated by biological lagoon and advanced oxidation processes. *J. Environ. Chem. Eng.*, 5, 6188–6193, 2017.
42. Ye, Z., Zhang, H., Zhang, X., Zhou, D., Treatment of landfill leachate using electrochemically assisted UV/chlorine process: effect of operating conditions, molecular weight distribution and fluorescence EEM-PARAFAC analysis. *Chem. Eng. J.*, 286, 508–516, 2016.
43. Zhao, X., Qu, J., Liu, H., Wang, C., Xiao, S., Liu, R., Liu, P., Lan, H., Hu, C., Photoelectrochemical treatment of landfill leachate in a continuous flow reactor. *Bioresour. Technol.*, 101, 865–869, 2010.
44. Panizza, M., Fine chemical industry, pulp and paper industry, petrochemical industry and pharmaceutical industry, in: *Electrochemical Water and Wastewater Treatment*, pp. 335–364, Elsevier Inc, 2018.
45. Lalwani, J., Gupta, A., Thatikonda, S., Subrahmanyam, C., An industrial insight on treatment strategies of the pharmaceutical industry effluent with varying qualitative characteristics. *J. Environ. Chem. Eng.*, 8, 104190, 2020.
46. Elsayed, A.M., Mahmoud, M., Abdel Karim, G.S.A., Abdelraof, M., Othman, A.M., Purification and biochemical characterization of two laccase isoenzymes isolated from *Trichoderma harzianum* S7113 and its application for bisphenol A degradation. *Microb. Cell Fact.*, 22, 1, 1–12, 2023, 7.
47. Othman, A.M., Mahmoud, M., Abdel Karim, G.S.A., Abdelraof, M., Elsayed, A.M., Enhancement of laccase production from a newly isolated *Trichoderma harzianum* S7113 using submerged fermentation: Optimization of production medium via central composite design and its application for hydroquinone degradation. *Int. J. Biol. Macromol.*, 192, 219–231, 2021.
48. Xu, D., Xiao, Y., Pan, H., Mei, Y., Toxic effects of tetracycline and its degradation products on freshwater green algae. *Ecotoxicol. Environ. Saf.*, 174, 43–47, 2019.
49. Fang, T., Liao, L., Xu, X., Peng, J., Jing, Y., Removal of COD and colour in real pharmaceutical wastewater by photoelectrocatalytic oxidation method. *Environ. Technol.*, 34, 779–786, 2013.
50. de Vidales, M.J.M., Millán, M., Sáez, C., Cañizares, P., Rodrigo, M.A., Irradiated-assisted electrochemical processes for the removal of persistent pollutants from real wastewater. *Sep. Purif. Technol.*, 175, 428–434, 2017.
51. Dindaş, G.B., Çalışkan, Y., Çelebi, E.E., Tekbaş, M., Bektaş, N., Yatmaz, H.C., Treatment of pharmaceutical wastewater by combination of electrocoagulation,

electro-fenton and photocatalytic oxidation processes. *J. Environ. Chem. Eng.*, 8, 103777, 2020.

52. GilPavas, E., Dobrosz-Gómez, I., Gómez-García, M.Á., Optimization of solar-driven photo-electro-Fenton process for the treatment of textile industrial wastewater. *J. Water Process Eng.*, 24, 49–55, 2018.
53. Bharagava, R.N. and Mishra, S., Hexavalent chromium reduction potential of *Cellulosimicrobium* sp. isolated from common effluent treatment plant of tannery industries. *Ecotoxicol. Environ. Saf.*, 147, 102–109, 2018.
54. Khater, D.Z., Amin, R.S., Fetohi, A.E., Mahmoud, M., El-Khatib, K.M., Insights on hexavalent chromium(VI) remediation strategies in abiotic and biotic dual chamber microbial fuel cells: electrochemical, physical, and metagenomics characterizations. *Sci. Rep.*, 13, 20184, 2023.
55. Lofrano, G., Meriç, S., Zengin, G.E., Orhon, D., Chemical and biological treatment technologies for leather tannery chemicals and wastewaters: a review. *Sci. Total Environ.*, 461/462, 265–281, 2013.
56. Borba, F.H., Seibert, D., Pellenz, L., Espinoza-Quiñones, F.R., Borba, C.E., Módenes, A.N., Bergamasco, R., Desirability function applied to the optimization of the Photoperoxi-Electrocoagulation process conditions in the treatment of tannery industrial wastewater. *J. Water Process Eng.*, 23, 207–216, 2018.
57. Selvaraj, H., Aravind, P., George, H.S., Sundaram, M., Removal of sulfide and recycling of recovered product from tannery lime wastewater using photoassisted-electrochemical oxidation process. *J. Ind. Eng. Chem.*, 83, 164–172, 2020.
58. Isarain-Chávez, E., De La Rosa, C., Godínez, L.A., Brillas, E., Peralta-Hernández, J.M., Comparative study of electrochemical water treatment processes for a tannery wastewater effluent. *J. Electroanal. Chem.*, 713, 62–69, 2014.
59. Ma, X., Zhang, J., Wang, B., Li, Q., Chu, S., Hierarchical Cu_2O foam/g-C_3N_4 photocathode for photoelectrochemical hydrogen production. *Appl. Surf. Sci.*, 427, 907–916, 2018.
60. Mahmoud, M., Parameswaran, P., Torres, C.I., Rittmann, B.E., Electrochemical techniques reveal that total ammonium stress increases electron flow to anode respiration in mixed-species bacterial anode biofilms. *Biotechnol. Bioeng.*, 114, 6, 1151–1159, 2017.
61. Mahmoud, M. and El-Khatib, K.M., Three-dimensional graphitic mesoporous carbon-doped carbon felt bioanodes enables high electric current production in microbial fuel cells. *Int. J. Hydrogen Energy*, 45, 56, 32413–32422, 2020.
62. Khater, D.Z., Amin, R.S., Zhran, M.O., Abd El-Aziz, Z.K., Mahmoud, M., Hassan, H.M., El-Khatib, K.M., The enhancement of microbial fuel cell performance by anodic bacterial community adaptation and cathodic mixed nickel–copper oxides on a graphene electrocatalyst. *J. Genet. Eng. Biotechnol.*, 20, 12, 2022.

63. Mahmoud, M., Torres, C.I., Rittmann, B.E., Changes in glucose fermentation pathways as a response to the free ammonia concentration in microbial electrolysis cells. *Environ. Sci. Technol.*, 51, 22, 13461–13470, 2017.
64. Lu, X., Xie, S., Yang, H., Tong, Y., Ji, H., Photoelectrochemical hydrogen production from biomass derivatives and water. *Chem. Soc. Rev.*, 43, 7581–7593, 2014.
65. Mahmoud, M., El-Kalliny, A.S., Squadrito, G., Stacked titanium dioxide nanotubes photoanode facilitates unbiased hydrogen production in a solar-driven photoelectrochemical cell powered with a microbial fuel cell treating animal manure wastewater. *Energy Convers. Manag.*, 254, 43, 115225, 2022.
66. Ishihara, H., Bock, J.P., Sharma, R., Hardcastle, F., Kannarpady, G.K., Mazumder, M.K., Biris, A.S., Electrochemical synthesis of titania nanostructural arrays and their surface modification for enhanced photoelectrochemical hydrogen production. *Chem. Phys. Lett.*, 489, 81–85, 2010.
67. Chehade, G., Alrawahi, N., Yuzer, B., Dincer, I., A photoelectrochemical system for hydrogen and chlorine production from industrial waste acids. *Sci. Total Environ.*, 712, 136358, 2020.
68. Chen, Z., Jaramillo, T.F., Deutsch, T.G., Kleiman-Shwarsctein, A., Forman, A.J., Gaillard, N., Garland, R., Takanabe, K., Heske, C., Sunkara, M., McFarland, E.W., Domen, K., Milled, E.L., Dinh, H.N., Accelerating materials development for photoelectrochemical hydrogen production: standards for methods, definitions, and reporting protocols. *J. Mater. Res.*, 25, 3–16, 2010.
69. Raptis, D., Dracopoulos, V., Lianos, P., Renewable energy production by photoelectrochemical oxidation of organic wastes using WO_3 photoanodes. *J. Hazard. Mater.*, 333, 259–264, 2017.
70. Saraswat, S.K., Rodene, D.D., Gupta, R.B., Recent advancements in semiconductor materials for photoelectrochemical water splitting for hydrogen production using visible light. *Renew. Sustain. Energy Rev.*, 89, 228–248, 2018.
71. Baltrusaitis, J., Hu, Y., Mcfarland, E.W., Hellman, A., Photoelectrochemical hydrogen production on a-Fe_2O_3 (0001): insights from theory and experiments. *ChemSusChem*, 3, 162–171, 2014.
72. Demir, M.E., Chehade, G., Dincer, I., Yuzer, B., Selcuk, H., Synergistic effects of advanced oxidization reactions in a combination of TiO_2 photocatalysis for hydrogen production and wastewater treatment applications. *Int. J. Hydrogen Energy*, 44, 23856–23867, 2019.
73. Gutierrez, M.I.J., Revero, E.P., Diaz, M.R.C., Gomez, M.E.N., Avella, J.A.P., Photoelectrocatalytic hydrogen production from oilfield-produced wastewater in a filter-press reactor using TiO_2-based photoanodes. *Catal. Today*, 266, 17–26, 2016.
74. Khater, D.Z., Amin, R.S., Fetohi, A.E., Mahmoud, M., El-Khatib, K.M., Reduced graphene oxide-supported palladium oxide-MOx for improving the performance of air-cathode microbial fuel cells: Influence of the Sn, Ce, Zn, and Fe precursors. *J. Power Sources*, 591, 233809, 2024.

75. Khater, D.Z., Amin, R.S., Mahmoud, M., El-Khatib, K.M., Evaluation of mixed transition metal (Co, Mn, and Cu) oxide electrocatalysts anchored on different carbon supports for robust oxygen reduction reaction in neutral media. *RSC Adv.*, 12, 2207–2218, 2022.
76. Khater, D.Z., Amin, R.S., Fetohi, A.E., El-Khatib, K.M., Mahmoud, M., Bifunctional manganese oxide–silver nanocomposites anchored on graphitic mesoporous carbon to promote oxygen reduction and inhibit cathodic biofilm growth for long-term operation of microbial fuel cells fed with sewage. *Sustain. Energy Fuels*, 6, 2, 430–439, 2022.
77. Chen, H., Chen, K., Lai, S., Dang, Z., Peng, Y., Photoelectrochemical oxidation of azo dye and generation of hydrogen via C-N co-doped TiO_2 nanotube arrays. *Sep. Purif. Technol.*, 146, 143–153, 2015.
78. Zhao, X., Pan, D., Chen, X., Li, R., Jiang, T., Wang, W., Li, G., Leung, D.Y.C., g-C_3N_4 photoanode for photoelectrocatalytic synergistic pollutant degradation and hydrogen evolution. *Appl. Surf. Sci.*, 467/468, 658–665, 2019.
79. Guaraldo, T.T., Gonçales, V.R., Silva, B.F., De Torresi, S.I.C., Zanoni, M.V.B.B., De Torresi, S.I.C., Zanoni, M.V.B.B., Hydrogen production and simultaneous photoelectrocatalytic pollutant oxidation using a TiO_2/WO_3 nanostructured photoanode under visible light irradiation. *J. Electroanal. Chem.*, 765, 188–196, 2016.
80. Iervolino, G., Tantis, I., Sygellou, L., Vaiano, V., Sannino, D., Lianos, P., Photocurrent increase by metal modification of Fe_2O_3 photoanodes and its effect on photoelectrocatalytic hydrogen production by degradation of organic substances. *Appl. Surf. Sci.*, 400, 176–183, 2017.
81. Wang, D., Li, Y., Li Puma, G., Wang, C., Wang, P., Zhang, W., Wang, Q., Dye-sensitized photoelectrochemical cell on plasmonic Ag/AgCl @ chiral TiO_2 nanofibers for treatment of urban wastewater effluents, with simultaneous production of hydrogen and electricity. *Appl. Catal. B: Environ.*, 168/169, 25–32, 2015.
82. Koo, M.S., Cho, K., Yoon, J., Choi, W., Photoelectrochemical degradation of organic compounds coupled with molecular hydrogen generation using electrochromic TiO_2 nanotube arrays. *Environ. Sci. Technol.*, 51, 6590–6598, 2017.
83. Zhou, Z., Wu, Z., Xu, Q., Zhao, G., A solar-charged photoelectrochemical wastewater fuel cell for efficient and sustainable hydrogen production. *J. Mater. Chem.*, 5, 25450–25459, 2017.

Index

Also of Interest

Check out these other books by Elvis Fosso-Kankeu published by Scrivener Publishing

Synergy of Bio-Chemical Processes for Photocatalytic and Photoelectrochemical Wastewater Treatment
Edited by Sadanand Pandey, Elvis Fosso-Kankeu and Soumya Pandit
Published 2024. ISBN 9781394197873

Recovery of Values from Low-Grade and Complex Minerals Development of Sustainable Processes
Edited by Elvis Fosso-Kankeu, Bhekie B. Mamba and Antoine F. Mulaba-Bafubiandi
Published 2024. ISBN 9781119896418

Hybridized Technologies for the Treatment of Mining Effluents
Edited by Elvis Fosso-Kankeu and Bhekie Mamba
Published 2023. ISBN 978-1-119-89642-5

Photoreactors in Advanced Oxidation Processes The Future of Wastewater Treatment
Edited by Elvis Fosso-Kankeu, Sadanand Pandey, Suprakas Sinha Ray
Published 2023. ISBN 978-1-394-16629-9

Photocatalysts in Advanced Oxidation Processes for Wastewater Treatment
Elvis Fosso-Kankeu, Sadanand Pandey and Suprakas Sinha Ray
Published 2020. ISBN 978-1-119-63139-2

Recovery of Byproducts from Acid Mine Drainage Treatment
Edited by Elvis Fosso-Kankeu, Christian Wolkersdorfer and Jo Burgess
Published 2020. ISBN 978-1-119-62007-5

Nano and Bio-Based Technologies for Wastewater Treatment
Prediction and Control Tools for the Dispersion of Pollutants in the Environment
Edited by Elvis Fosso-Kankeu
Published 2019 ISBN 978-1-119-57709-6

www.Scrivenerpublishing.com

Printed in the USA/Agawam, MA
September 16, 2026

910661.065